# EPIOPTICS-11

# THE SCIENCE AND CULTURE SERIES — PHYSICS

*Series Editor:* A. Zichichi, *European Physical Society, Geneva, Switzerland*

*Series Editorial Board:* P. G. Bergmann, J. Collinge, V. Hughes, N. Kurti, T. D. Lee, K. M. B. Siegbahn, G. 't Hooft, P. Toubert, E. Velikhov, G. Veneziano, G. Zhou

THE SCIENCE AND CULTURE SERIES — PHYSICS

# EPIOPTICS-11

Proceedings of the 49th Course of the
International School of Solid State Physics

Erice, Italy   19 — 25 July 2010

Editors
**Antonio Cricenti**

Series Editor
**A. Zichichi**

**World Scientific**

NEW JERSEY · LONDON · SINGAPORE · BEIJING · SHANGHAI · HONG KONG · TAIPEI · CHENNAI

*Published by*

World Scientific Publishing Co. Pte. Ltd.

5 Toh Tuck Link, Singapore 596224

*USA office:* 27 Warren Street, Suite 401-402, Hackensack, NJ 07601

*UK office:* 57 Shelton Street, Covent Garden, London WC2H 9HE

**British Library Cataloguing-in-Publication Data**
A catalogue record for this book is available from the British Library.

**The Science and Culture Series — Physics**
**EPIOPTICS-11**
**Proceedings of the 49th Course of the International School of Solid State Physics**

Copyright © 2012 by World Scientific Publishing Co. Pte. Ltd.

ISBN 978-981-4417-11-2

Printed in Singapore by World Scientific Printers.

# PREFACE

This special World Scientific volume contains the Proceedings of the 11th Epioptics Workshop/School, held at the Ettore Majorana Foundation and Centre for Scientific Culture, Erice, Sicily, from July 19 to 25, 2010. This was the 11th Workshop/School in the Epioptics series and the 49th of the International School of Solid State Physics. Antonio Cricenti from CNR Istituto di Struttura della Materia and Theo Rasing from the University of Njimegen, were the Directors of the Workshop/School. The Advisory Committee of the Workshop included Y. Borensztein from U. Paris VII (F), R. Del Sole from U. Roma II Tor Vergata (I), D. Aspnes from NCSU (USA), O. Hunderi from U. Trondheim (N), J. McGilp from Trinity College Dublin (Eire), W. Richter from TU Berlin (D), N. Tolk from Vanderbilt University (USA), and P. Weightman from Liverpool University (UK). Forty-three scientists from 12 countries attended the workshop.

The workshop brought together researchers from universities and research institutes who work in the fields of (semiconductor) surface science, epitaxial growth, materials deposition and optical diagnostics relevant to (semiconductor) materials and structures of interest for present and anticipated (spin) electronic devices. The workshop was aimed at assessing the capabilities of state-of-the-art optical techniques in elucidating the fundamental electronic and structural properties of semiconductor and metal surfaces, interfaces, thin layers, and layer structures, and assessing the usefulness of these techniques for optimization of high quality multilayer samples through feedback control during materials growth and processing. Particular emphasis was dedicated to the theory of non-linear optics and to dynamical processes through the use of pump-probe techniques together with the search for new optical sources. Some new applications of scanning probe microscopy to material science and biological samples, dried and *in vivo*, with the use of different laser sources were also presented. Materials of particular interest were silicon, semiconductor-metal interfaces, semiconductor and magnetic multi-layers and III-V compound semiconductors. As well as the notes collected in this Volume, the Workshop combined tutorial aspects proper to a school with some of the most advanced topics in the field, which better characterized the workshop.

This book is dedicated to Wolfgang Richter (indicated by the white arrow in the picture during Epioptics-10), an outstanding scientist, in occasion of his 70 years birthday. It is a collection of articles that were presented at the 11th Epioptics Workshop/School – that was also in the honor of Wolfgang and attracted some of his friends worldwide. Wolfgang Richter, in a long and very distinguished career, made fundamental contributions to the development of Raman Spectroscopy and to its application in materials science.

I wish to thank Prof. A. Zichichi, President of the Ettore Majorana Foundation and Centre for Scientific Culture (EMFCSC), the Italian National Research Council (CNR) and the Sicilian Regional Government. I wish to thank Prof. G. Benedek, Director of the International School of Solid State Physics of the EMFCSC. Our thanks are also due to the Director for Administration and Organizational Affairs, Ms. F. Ruggiu and all the staff of the centre for their excellent work.

*Antonio Cricenti*

# CONTENTS

# EPITAXIAL-LIKE GROWTH OF LEAD PHTHALOCYANINE LAYERS ON GAAS(001) SURFACES

L. Riele[a], B. Buick[a], E. Speiser[b], B.-O. Fimland[c], P. Vogt[d], W. Richter[a]

[a] Dipartimento di Fisica, Università di Roma Tor Vergata, Via della Ricerca Scientifica 1, I-00133 Rome, Italy,
[b] Leibniz-Institut für Analytische Wissenschaften - ISAS - e.V., Albert-Einstein-Str.9, D-12489 Berlin, Germany,
[c] Department of Electronics and Telecommunications, Norwegian University of Science and Technology, NO-7491 Trondheim, Norway,
[d] Institut für Festkörperphysik, Technische Universität Berlin, Hardenbergstr.36, D-10623 Berlin, Germany

Lead phthalocyanine (PbPc) layers were deposited on GaAs(001)-$c(4 \times 4)$ and -$(2 \times 4)$ reconstructed surfaces and studied with the goal to determine a possible orientation with respect to the substrate in the layers by Raman spectroscopy. Moreover, the dependence of the molecular arrangement on the GaAs surface reconstruction was tested. For this purpose the intensity of the Raman peaks was measured as a function of a rotation of the sample around its surface normal. Since every molecule has a specific Raman tensor attached to its internal coordinates any order of the molecules in the layer should be indicated by specific variations in intensity with the rotation, except in case of a completely statistical arrangement (amorphous structure). Indeed periodic changes with rotation of the measured intensities of the molecular vibrational modes were observed. They are compared to the results of calculations with the corresponding Raman tensors and with the GaAs(001) phonon modes. The periodicity of the measured changes and their fixed relation to the substrate coordinates suggest well-ordered layers with preferential molecular orientations. Together with first observations of different structural layer properties on the -$c(4 \times 4)$ and -$(2 \times 4)$ surfaces this implies that the molecular orientation is induced by the atomic arrangement on the GaAs surfaces. Thus, we conclude that an epitaxial-like growth mode of PbPc molecules on reconstructed GaAs(001) surfaces takes place. This is an important result for organic devices like OLEDS or OFETS.

Keywords: lead phthalocyanine, GaAs(001), Raman spectroscopy, organic MBE, molecular orientation

## 1. Introduction

Heterostructures of semiconductors and metal phthalocyanine layers have attracted growing scientific and industrial interest due to their semiconducting properties and consequently potential applications in optoelectronic or electronic devices.[1] Owing to the wide flexibility in terms of their electronic properties by easily modifying their composition and structure, these organic components are possible candidates to replace inorganic materials. Potential applications include organic field-effect transistors (OFETs),[2] organic light-emitting diodes (OLEDs, displays),[3] organic solar cells[4] and gas-sensors.[5,6] An important aspect in this respect is the direction of the charge transport. For OFETs the charge transport must be oriented parallel to the substrate surface and for OLEDs and organic solar cells perpendicular. In organic layers the maximum electrical conductivity depends crucially on the orientation and ordering of the molecules.[7] Hence, depending on the required direction of the electrical conductivity within the organic layer the orientation of the molecules has to be adjusted and ordered organic layers on inorganic semiconductor surfaces are therefore needed for the production of electronic devices. In the case of lead phthalocyanine (PbPc) molecules the highest conductivity is perpendicular to the molecular plane as indicated in Fig. 1.[8,9] Fig. 1 illustrates the ordering of PbPc molecules for OFETs (standing molecules) and OLEDs (lying molecules) to obtain the intended charge transport. Phthalocyanine is an organic ringmolecule composed of four isoindole groups forming a macrocycle with a central metal atom (Fig. 2). For PbPc, in particular, the $Pb^{2+}$-ion radius is too large to fit into the macrocycle. This leads to an out-of-plane position of the $Pb^{2+}$-ion and gives rise to a dome-like structure (Fig. 2).

Amongst metal Pcs the most frequently investigated are the planar copper phthalocyanine (CuPc)[15-21] as well as the non-planar tin phthalocyanine (SnPc)[22-24] and lead phthalocyanine (PbPc).[25-28] The growth of PbPc on different reconstructed substrate surfaces has already been investigated with several surface sensitive techniques, based on their electronic properties and mostly focused on the interface formation within the first monolayer.[26-28,52] A good control on the degree of order of molecular layers on semiconductor surfaces has been demonstrated by controlling several growth parameters, such as the substrate temperature, evaporation rate, and post-evaporation annealing.[29] During the last years the influence of the substrate surface on the molecular ordering[11,13,30-33] has become more and more the center of attention.

In the epitaxial growth of group IV and III-V semiconductors the growth process consists of single atoms bonding (covalently) to substrate surfaces and the materials have at least similar lattice parameters.[10] The epitaxial growth of organic molecules on inorganic substrates is more complicated.

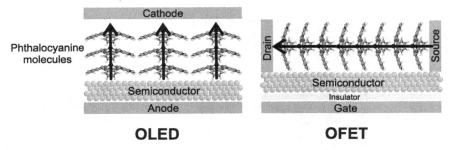

Fig. 1. Electrical conductivity depends on the orientation of the molecules. An application for lying phthalocyanine molecules is e.g. OLEDs (left) and for standing OFETs (right). The arrows indicate the direction of the highest electrical conductivity.[8]

The growth process consists of whole molecules bonding to or interacting "at once" with the substrate surface, thus the lattice parameters are very different and the bonding mechanism more complicated. Nevertheless, it has been shown that it is possible to deposit well-ordered molecular layers in an epitaxial-like manner, which means that the crystalline structure within the organic layers can be determined by the substrate surfaces.[11,13] However, since the forces within organic crystals are mainly due to the rather weak van der Waals forces a relaxation of the organic structure into the corresponding bulk structure within a few monolayers[11] is very likely. Therefore, in general, only the orientation within the first monolayers can be controlled by the surface and not the entire morphology of the organic layer.[12]

During the last years Raman spectroscopy has become an important method to investigate the growth and the structural properties of organic layers from several monolayers up to thicker layers.[35] Vibrational modes have characteristic frequencies and symmetries. Hence, the study of said vibrational modes provides information about the orientation of molecules in organic layers. An extensive summary of possible applications of Raman spectroscopy to study organic molecular layers has been given by Zahn *et al.*[35]

Here, we present Raman spectroscopy investigations of PbPc layers deposited mainly on the GaAs(001)-$c(4 \times 4)$ surface reconstruction.[36–39] These are compared to first spectra of layers on GaAs(001)-$(2 \times 4)$. Prior investigations of the adsorption and layer growth of PbPc on the GaAs(001)-$c(4 \times 4)$ and $(2 \times 4)$ surface reconstructions from sub-monolayers up to 20 nm thick layers by scanning tunneling microscopy and reflectance anisotropy spectroscopy (RAS) have already been carried out.[52] It was shown that the

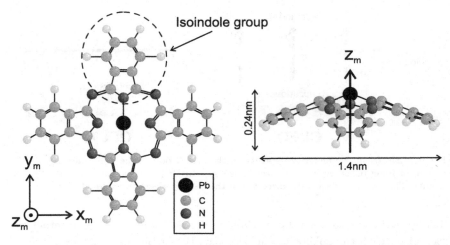

Fig. 2.   Structure model (top and side view) of PbPc after Papageorgiou *et al.*[34] The molecular coordinate axes $x_m$ and $y_m$ are parallel to the perpendicular isoindole groups and $z_m$ corresponds to the 4-fold rotational axis $C_4$.

adsorption geometry of the first monolayer depends on the atomic ordering of the substrate surface which induces different epitaxial-like growth modes of thicker PbPc layers on GaAs(001). RAS revealed that the ordering is maintained within the thicker layers but does not give evidence to the orientation of the molecules within these ordered structures. The determination of, among other things, the azimuthal orientation of the molecules can be solved by Raman spectroscopy measurements. The aim is to determine the orientation of the molecules with respect to the substrate within thicker layers. For that purpose the Raman scattering intensity is measured as a function of the rotation of the sample around its normal and compared to the calculated results of a rotation of the corresponding Raman tensors.

## 2. Experimental

### 2.1. *Sample Preparation*

The samples were prepared under ultra-high vacuum (UHV) conditions (base pressure $< 2 \times 10^{-10}$ mbar). The GaAs(001) substrates were grown by molecular beam epitaxy (MBE) and are doped with Si (n = $5 \times 10^{17}$ cm$^{-3}$). A capping layer of amorphous Arsenic protects the surface for a contamination-free transfer through air.[40,41] The GaAs(001)-$c(4 \times 4)$ surface reconstruction was obtained by thermal desorption of the protection layer at 350°C ($\pm 20$°C) inside the UHV chamber.[41,42] Further annealing up to 430°C (±20°C) is leading to the (2 × 4) reconstruction. During the growth of the organic layer the samples were kept at room temperature. For the deposition of lead phthalocyanine (PbPc) on GaAs(001)-c(4×4) and (2 × 4) a well degassed water cooled Knudsen cell was kept at a temperature of about 350°C (±20°C).

## 2.2. Raman Spectroscopy

Raman spectroscopy deals with the analysis of inelastically scattered light from solids or molecules. During the interaction of light and matter excitation and annihilation of phonons are possible, leading to a change in frequency of the incident light. In a classical approach Raman scattering is described by a modulation of the polarizability of molecules through their vibrations resulting in scattered electromagnetic fields with shifted frequencies from the frequency of the incident field. The change in polarizability due to a molecular vibration is called Raman tensor $R_{(vib)}$.

The Raman scattering intensity $I_s$ is proportional to[43,44]

$$I_s \propto |\vec{e}_s \cdot R_{(vib)} \cdot \vec{e}_i|^2 \tag{1}$$

where $\vec{e}_s$ and $\vec{e}_i$ are the unit vectors of the incident and scattered light and $R_{(vib)}$ is the Raman tensor of a specific vibrational mode.

Raman polarization studies have been utilized to determine the order and orientation of molecules within layers composed of organic molecules.[45] The symmetry of the vibrations may be described with the help of group theory. Each vibration exhibits a symmetry corresponding to one of the irreducible representations (Raman tensors) of the corresponding symmetry group. In the case of PbPc the $C_{4v}$ point group. The symmetries of the Raman active molecular vibrations of a PbPc molecule are represented by the following Raman tensors described in molecular coordinates:

$$R(A_1) = \begin{pmatrix} a & 0 & 0 \\ 0 & a & 0 \\ 0 & 0 & b \end{pmatrix} \quad R(B_1) = \begin{pmatrix} c & 0 & 0 \\ 0 & -c & 0 \\ 0 & 0 & 0 \end{pmatrix} \quad R(B_2) = \begin{pmatrix} 0 & d & 0 \\ d & 0 & 0 \\ 0 & 0 & 0 \end{pmatrix} \tag{2}$$

$$R(E, x) = \begin{pmatrix} 0 & 0 & e \\ 0 & 0 & 0 \\ e & 0 & 0 \end{pmatrix} \quad R(E, y) = \begin{pmatrix} 0 & 0 & 0 \\ 0 & 0 & e \\ 0 & e & 0 \end{pmatrix} \tag{3}$$

where a, b, c, d and e correspond to the change in the polarizability.

A molecular layer, containing equally oriented molecules, can be macroscopically described by the Raman tensors for a single molecule by averaging over many. Hence, measuring the Raman scattering intensity as a function of the rotation of the sample provides information about the orientation of the molecules within layers. Disordered (amorphous) molecular layers would show on the other hand no intensity changes. The degree of order can be derived from the relative strength of the intensity changes with respect to a constant "underground" due to disordered molecules.

The aim is to determine ordering and the orientation of the molecules with respect to the GaAs substrate. For the investigated samples the molecular layers were sufficiently thin to observe vibrations originating from molecules and GaAs substrate at the same time. Since for GaAs the change of Raman scattering intensity in correlation to the crystalline axes is known it can thus serve as a reference system.

In the GaAs crystal for the scattering geometries $\vec{e}_s \parallel \vec{e}_i$ and $\vec{e}_s \perp \vec{e}_i$ (where $\vec{e}_i$ is parallel and $\vec{e}_s$ either parallel or perpendicular to the [0$\bar{1}$1] crystallographic direction of GaAs(001)) only Raman scattering by the LO phonon is allowed. If the experimental geometry is $\vec{e}_s \parallel \vec{e}_i$ it is referred to as *parallel* and for $\vec{e}_s \perp \vec{e}_i$ *crossed* polarization configuration. The Raman tensor of the GaAs LO mode is

$$R(GaAs_{LO}) = \begin{pmatrix} 0 & f & 0 \\ f & 0 & 0 \\ 0 & 0 & 0 \end{pmatrix}, \tag{4}$$

where f corresponds to the the change in the susceptibility.

A method to determine the molecular orientation by Raman scattering based on the rotation of the sample is described in.[35,47] There, two coordinate transformations are applied to the Raman tensors in order to describe rotational dependencies of the Raman scattering intensity. One transforms the molecular to the reference (here: substrate) coordinate system introducing the Euler angles $(\varphi, \theta, \psi)$ to relate the molecular orientation to the substrate. $\theta$ is the tilt angle of the molecule with respect to the substrate surface normal z and $\psi$ the rotation of the molecule around the z-axis. The

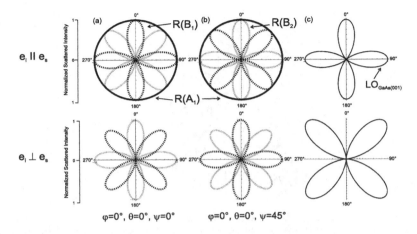

Fig. 3. Intensity changes over 360° upon transformation of the Raman tensors of the $C_{4v}$ point group for different Euler angle configurations and for the GaAs LO phonon. All axes show the intensity in arb. units. In this work only the azimuthal orientation is taken into account.

third Euler angle is set to $\varphi = 0$, since we expect a combination of tilt and rotation. The other transformation describes the rotation around one axis. In principle, rotations around every axis would be possible. For experimental simplicity a rotation around the samples normal ($\gamma$) was chosen. The Raman scattering intensity in Eq. (1) depends then on the transformed Raman tensors $R_{rot}(\varphi, \theta, \psi, \gamma)$.

The Raman scattering intensity, as a function of the rotation of the sample, can then be calculated for different molecular orientations. Fig. 3 shows the intensity variations for two different molecular orientations and for the GaAs substrate in the $\vec{e}_s \parallel \vec{e}_i$ (top) and $\vec{e}_s \perp \vec{e}_i$ (bottom) configuration, respectively. The columns (a) and (b) in Fig. 3 illustrate different combinations of tilt and rotation for the Raman tensors $R(B_1)$, $R(B_2)$, and $R(A_1)$. The intensity variation of the GaAs LO phonon, as a function of the rotation around z (corresponds to [001]), is shown in column (c). The rotation of the molecule of $\psi = 45°$ causes a shift of the maxima of the two Raman tensors $R(B_1)$ and $R(B_2)$. $R(A_1)$, however, stays constant for both cases in the $\vec{e}_s \parallel \vec{e}_i$ configuration and disappears in the $\vec{e}_s \perp \vec{e}_i$ configuration.

During the experiment samples of PbPc layers deposited on GaAs(001)-$c(4 \times 4)$ were rotated clockwise in steps of $\Delta\gamma = 15°$. The measured Raman scattering intensity was thus measured as a function of the angle $\gamma$ between

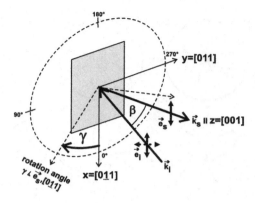

Fig. 4. For the analysis of the rotational dependence of the Raman scattering intensity of the molecular vibrational modes the sample is rotated clockwise in steps of 15° around its normal. At an angle $\gamma = 0°$ the electric field vector is parallel to the [0$\bar{1}$1] crystalline axis of the GaAs(001) substrate.

x= [0$\bar{1}$1] and $\vec{e}_s$. Another aspect is the investigation of the polarization dependence of the Raman scattering intensity in the different scattering geometries. This is very useful to determine the symmetry (Raman tensor) of Raman-active molecular vibrational modes[49] and to be able to assign vibrational modes to their corresponding Raman tensor because of their different behavior.

The coordinates of the laboratory system (x,y,z) were assigned as followed: The x and y axis are defined to be parallel to the two perpendicular crystalline axes of the GaAs(001) substrates (x= [0$\bar{1}$1], y= [011], z= [001] the surface normal) (Fig. 4). The polarization of the incident radiation (laser) was aligned either parallel or perpendicular to x while the scattered radiation was always observed parallel to z, corresponding to $\vec{e}_s \parallel \vec{e}_i$ and $\vec{e}_s \perp \vec{e}_i$ configuration, respectively. At an angle $\gamma = 0°$ the electric field vector $\vec{e}_s$ was parallel to the [0$\bar{1}$1] crystalline axis of the GaAs(001) substrate. The angle between $\vec{k}_i$ and the z axis was $\beta = 40°$ (Fig. 4). The rotation was done for the $\vec{e}_s \parallel \vec{e}_i$ and $\vec{e}_s \perp \vec{e}_i$ configuration. The excitation source is an Ar$^+$ laser. Further details on the Raman set-up can be found in.[50]

## 3. Results

Fig. 5 shows an overview of the Raman spectra in the spectral range $200 - 3100\,cm^{-1}$ of a PbPc layer ($< 50nm$) deposited on reconstructed GaAs(001) surfaces. The spectra were taken in ambient conditions. Fig. 5(a)

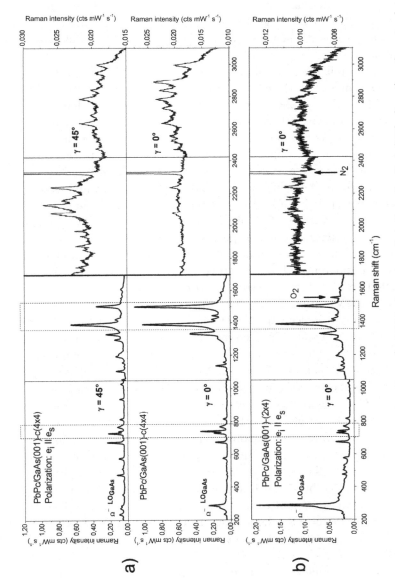

Fig. 5. Raman spectra in the range $200 - 3100$ cm$^{-1}$ of PbPc layer deposited a) on the GaAs(001)-$c(4 \times 4)$ surface at a low power density $20\,kW/cm^2$ corresponding to a laser power of $220\,mW$. Marked are molecular vibrational modes where the intensity ratio changes upon rotation of the sample around the angle $\gamma$ between x= $[0\bar{1}1]$ and $\bar{e}_s$. b) on the GaAs(001)-$(2 \times 4)$ surface with a power density of $\approx 16\,kW/cm^2$ corresponding to a laser power of $170\,mW$. The excitation energy was 2.54 eV (488 nm) lying in the Q absorption band of PbPc[53]

shows spectra taken at $\gamma = 0°$ (middle) and $45°$ (top) on the GaAs(001)-$c(4 \times 4)$ surface. At about $1556\,\mathrm{cm}^{-1}$ the O-O ($O_2$) stretching mode is observed.[43] This vibrational mode was not observed in vacuum and is therefore an indicator of the degree of oxidation due to exposure to air. Its intensity has increased when ex-situ measurements were repeated after several months.

The spectra exhibit peaks attributed to PbPc and the GaAs substrate, namely the LO phonon at $292\,\mathrm{cm}^{-1}$ of GaAs(001). As discussed above the observation of the $LO_{GaAs}$ phonon offers a reference system to determine the molecular orientation with respect to the GaAs substrate. Clearly a change of the intensity ratio between several modes is found for different $\gamma$. This is found e.g. in the range $700 - 750\,\mathrm{cm}^{-1}$ which can be assigned according to[51] to the macrocycle ring stretch and in the range $1350 - 1550\,\mathrm{cm}^{-1}$, assigned to the isoindole ring and the C-N aza-group stretch. Between $2800 - 3000\,\mathrm{cm}^{-1}$ the C-H stretching mode region is found.

Fig. 5(b) shows a spectrum of a PbPc layer deposited on the GaAs(001)-$(2 \times 4)$ surface. A comparison of the spectra with each other shows differences in intensity ratio at $\gamma = 0°$ especially in the ranges discussed before. This implies that apparently the structural properties within the molecular layers deposited on two different reconstructions are different.

To illustrate the variations in intensity Fig. 6(a) shows the Raman scattering intensity of the molecular vibrational modes as a function of the rotation of $360°$ with $\Delta\gamma = 15°$c. The measurements were done in $\vec{e}_s \parallel \vec{e}_i$ and $\vec{e}_s \perp \vec{e}_i$ configuration. In order to illustrate the relative orientation of the molecular vibrational modes to the GaAs(001) LO mode the range $200 - 1070\,\mathrm{cm}^{-1}$ is shown. As a function of the rotation the intensity of the LO phonons of GaAs as well as the intensity of most molecular vibrations vary with a periodicity of $90°$ in both $\vec{e}_s \parallel \vec{e}_i$ and $\vec{e}_s \perp \vec{e}_i$ configurations. For the $\vec{e}_s \parallel \vec{e}_i$ configuration the maximum intensity of the LO phonons is found, as expected, when the polarization is parallel to the $[011]$ and $[0\bar{1}1]$ crystalline axes of the GaAs(001) substrate. In the $\vec{e}_s \perp \vec{e}_i$ configuration this periodicity is shifted by $45°$ with respect to the $\vec{e}_s \parallel \vec{e}_i$ configuration. As marked in the graphs some molecular vibrational modes, e.g. as indicated the $746\,\mathrm{cm}^{-1}$, of PbPc have the same phase as the GaAs phonons whereas others, e.g. as indicated $944\,\mathrm{cm}^{-1}$, occur $45°$ phase shifted to them. In the $\vec{e}_s \perp \vec{e}_i$ configuration the phase of the observed variations is also shifted by $45°$, respectively. Other vibrations, e.g. $820\,\mathrm{cm}^{-1}$, however, show a constant intensity upon rotation in the $\vec{e}_s \parallel \vec{e}_i$ configuration whereas they almost disappear in the $\vec{e}_s \perp \vec{e}_i$.

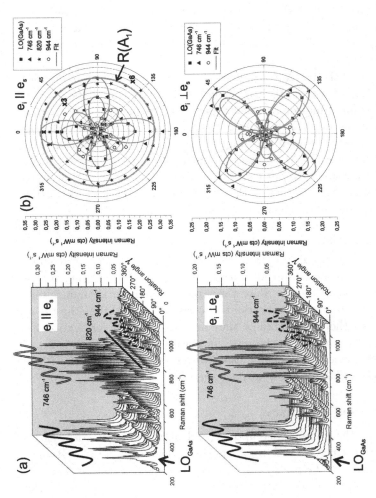

Fig. 6. Dependence of the intensity of a PbPc layers deposited on the GaAs(001)-$c(4 \times 4)$ surface upon rotation of the sample. The excitation laser line was 488 nm. (a) shows the spectral range $200 - 1050 \, cm^{-1}$ for $\vec{e}_s \parallel \vec{e}_i$ and $\vec{e}_s \perp \vec{e}_i$ configuration containing the LO phonon of GaAs(001). The "waves" with different linestyles represent the intensity changes of the differently behaving vibrational modes. Most vibrational modes, amongst others the LO phonon, vary in intensity by a periodicity of 90° during the rotation. Some vibrational modes on the contrary are constant. In the polarplots (b) the periodicities are shifted by 45° in $\vec{e}_s \parallel \vec{e}_i$ and $\vec{e}_s \perp \vec{e}_i$ configurations, respectively.

In Fig. 6(b) polarplots are shown for the marked modes in Fig. 6(a). The experimental data are fitted by the rotated Raman tensors $R_{rot}(\varphi, \theta, \psi, \gamma)$ in Eq. (1). In the case of the $\vec{e}_s \parallel \vec{e}_i$ configuration one molecular mode ($746\,\text{cm}^{-1}$) behaves similar to the $LO_{GaAs}$ phonon. Their maxima in intensity occur at $\gamma = 0°$ repeating every 90°. Another mode ($944\,\text{cm}^{-1}$) on the other hand occurs 45° shifted to them having a 90° periodicity, as well. An example for a constant mode ($820\,\text{cm}^{-1}$) is also given. As illustrated in the $\vec{e}_s \perp \vec{e}_i$ configurations all the phases are shifted by 45° and the constant mode does not appear.

The behavior of the intensity changes of the different vibrational modes seen in Fig. 6(b) are comparable to the calculated rotation of the Raman tensors (Fig. 3) for a molecular orientation either parallel or 45° rotated with respect to the crystalline axes of GaAs. The modes with constant intensity can be assigned with certainty to the $R(A_1)$ Raman tensor. Yet, for the vibrations with a 90° periodicity an assignment to either $R(B_1)$ or $R(B_2)$ is not possible without e.g. further lattice dynamics calculations. Without an assignment a precise statement about the molecular orientation can therefore not be given. Thus, two possible orientations of the molecules either parallel or $\psi = 45°$ rotated around the z axis with respect to the crystalline axes of GaAs are possible.

A tilting of the molecules ($\theta$) with respect to the substrate surface normal z would be depicted in an alternating change in the intensity of the maxima of one Raman tensor. Since the strength of the maximum intensity seems to be equal for all vibrational modes we assume that the tilt of the molecules is either very small or they are flat lying. A verification for normal incident is planned in the near future to exclude the influence of $\beta$. Furthermore, since the intensity vanishes almost completely the PbPc layers on the GaAs(001)-$c(4 \times 4)$ surface are very well ordered.

## 4. Conclusion and Summary

The rotation of the samples, composed of PbPc layers on GaAs(001)-$c(4 \times 4)$, around their normal showed a periodic dependence of the Raman scattering intensity of the molecular vibrational modes. This indicates ordering with a preferred orientation of the molecules with respect to the GaAs crystalline axes. We observed three different behaviors of intensity variations: one with a period correlated to the $LO_{GaAs}$ phonon, another with a period 45° shifted to them and a third type not changing in intensity upon rotation and disappearing in the $\vec{e}_s \perp \vec{e}_i$ configuration. This suggests that these molecular vibrational modes have different symmetries and thus different Raman tensors. Until now only the constant vibrational modes can be assigned clearly to the $R(A_1)$ Raman tensor. The other vibrational modes, having a 90° periodicity, have the same character then the $R(B_1)$ and $R(B_2)$ Raman tensors, however, a clear assignment is not yet possible. This will be solved by DFT calculations already in preparation which provide the needed frequencies and symmetries for a symmetry assignment of the molecular vibrational modes. Depending on the assignment the molecules are then either aligned with their isoindole group along the crystalline axes of the GaAs(001) substrate or rotated by 45°.

A first comparison to a spectrum of a PbPc layer deposited on the GaAs(001)-(2 × 4) surface indicates structural differences.

We can therefore conclude that molecules arrange with respect to the crystalline axes of the GaAs(001) substrates and that their ordering is induced by the surface reconstruction of the substrates.

## References

1. C. C. Leznoff and A. B. P. Lever, *Phthalocyanines - Properties and Applications* VCH, 1989, ISBN: 9780471188636.
2. Z. Bao, *Advanced Materials* **12**, 227 (2000).
3. D. Hohnholz, S. Steinbrecher and M. Hanack, *J. Mol. Struct* **521**, 231 (2000).
4. H. Yonehara and C. Pac, *Thin Solid Films* **278**, 108 (1996).
5. R. Zhou, F. Josse, W. Göpel, Z. Z. Öztürk and O. Bekaroglu, *Applied organometallic chemistry* **10**, 557 (1996).
6. A. Mrwa, M. Friedrich, A. Hofmann and D. Zahn, *Sensors and Actuators B: Chemical* **25**, 596 (1995).
7. K. Mizoguchi, K. Mizui, D. Kim and M. Nakayama, *Jpn. J. Appl. Phys.* **41**, 6421 (2002).
8. H. Yamane, H. Honda, H. Fukagawa, M. Ohyama, Y. Hinuma, S. Kera, K. K. Okudaira and N. Ueno, *Journal of Electron Spectroscopy and Related Phenomena* **137 - 140**, 223 (2004).
9. K. Ukei, K. Takamoto and E. Kanda, *Phys. Lett. A* **45**, 345 (1973).
10. M. A. Herman, W. Richter, H. Sitter, *Epitaxy: Physical Principles and Technical Implementation* (Springer-Verlag GmbH, 2004), ISBN: 3540678212.
11. S. R. Forrest, *Chem. Rev.* **97**, 1793 (1997).
12. N. B. McKeown, *Phthalocyanine materials* (Cambridge University Press, 1998), ISBN: 0521496233.
13. A. Koma, *Progress in Crystal Growth and Characterization of Materials* **30**, 129 (1995).
14. A. R. West, *Solid State Chemistry and Its Applications* (wiley, 1985).
15. P. H. Lippel, R. J. Wilson, M. D. Miller, C. Wöll and S. Chiang, *Phys. Rev. Lett.* **620**, 171 (1989).
16. P. H. Lippel, R. J. Wilson, M. D. Miller, C. Wöll and S. Chiang, *J. Vac. Sci. Technol. A* **8**, 659 (1990).

14

17. C. Ludwig, R. Strohmaier, J. Petersen, B. Gompf and W. Eisenmenger, *J. Vac. Sci. Technol. B* **12**, 1963 (1994).
18. M. Kanai, T. Kawai, K. Motai, X. Wang, T. Hashizume and T. Sakura, *Surf. Sci.* **329**, 619 (1995).
19. J. Y. Grand, T. Kunstmann, D. Hoffmann, A. Haas, M. Dietsche, J. Seifritz and R. Möller, *Surf. Sci.* **366**, 403 (1996).
20. M. Nakamura, Y. Morita, Y. Mori, A. Ishitani and H. Tokumoto, *J. Vac. Sci. Technol. B* **14**, 1109 (1996).
21. M. Stöhr, T. Wagner, M. Gabriel, B. Weyers and R. Möller, *Phys. Rev. B* **65**, 033404 (2001).
22. K. Walzer and M. Hietschold, *Surf. Sci.* **471**, 1 (2001).
23. M. Lackinger and M. Hietschold, *Surf. Sci. Lett.* **520**, L619 (2002).
24. E. Salomon, T. Angot, N. Papageorgiou and J. M. Layet, *Sur. Sci.* **596**, 74 (2005).
25. O. Pester, A. Mrwa and M. Hietschold, *phys. stat. sol. (a)* **131**, 19 (1992).
26. R. Strohmaier, C. Ludwig, J. Petersen, B. Gompf and W. Eisenmenger, *J. Vac. Sci. Technol. B* **14**, 1079 (1996).
27. L. Ottaviano, L. Lozzi, S. Santucci, S. D. Nardo and M. Passacantando, *Surf. Sci.* **392**, 52 (1997).
28. T. Angot, E. Salomon, N. Papageorgiou and J.-M. Layet, *Surf. Sci.* **572**, 59 (2004).
29. R. Collins and A. Belghachi, *Materials Letters* **8**, 349 (1989).
30. N. Papageorgiou, E. Salomon, T. Angot, J.-M. Layet, L. Giovanelli and G. Le Lay, *Prog. Surf. Sci.* **777**, 139 (2004).
31. J. Cox and T. Jones, *Surf. Sci.* **457**, 311 (2000).
32. C. Kendrick and A. Kahn, *Appl. Surf. Sci.* **123-124**, 405 (1998).
33. S. Yim and T. S. Jones, *Surf. Sci.* **521**, 151 (2002).
34. N. Papageorgiou, Y. Ferro, E. Salomon, A. Allouche, J. M. Layet, L. Giovanelli and G. Le Lay, *Phys. Rev. B* **68**, 235105 (2003).
35. D. R. T. Zahn, G. N. Gavrila and G. Salvan, *Chemical Reviews* **107**, 1161 (2007).
36. W. G. Schmidt, F. Bechstedt, W. Lu and J. Bernholc, *Phys. Rev. B* **66**, 085334 (2002).
37. W. G. Schmidt, F. Bechstedt and J. Bernholc, *Appl. Sur. Sci.* **190**, 264 (2002).
38. W. G. Schmidt, *Appl. Phys. A* **75**, 89 (2002).
39. N. Esser, W. G. Schmidt, C. Cobet, K. Fleischer, A. I. Shkrebtii, B. O. Fimland and W. Richter, *J. Vac. Sci. Technol. B* **19**, 1756 (2001).
40. U. Resch, W. Scholz, U. Rossow, A. B. Muller, W. Richter and A. Forster, *Appl. Surf. Sci.* **63**, 106 (1993).
41. R. W. Bernstein, A. Borg, H. Husby, B.-O. Finnland and J. K. Grepstad, *Appl. Surf Sci* **56-58**, 74 (1992).
42. U. Resch-Esser, N. Esser, D. T. Wang, M. Kuball, J. Zegenhagen, B. O. Fimland and W. Richter, *Surf. Sci.* **352-354**, 71 (1996).
43. D. A. Long, *The Raman Effect: A Unified Treatment of the Theory of Raman Scattering by Molecules* (John Wiley&Sons Ltd, 2002.

44. G. Turrell, *Raman Microscopy: Developments and Applications* (ACADEMIC PR INC, 1996.

45. T. Basova and B. Kolesov, *Thin Solid Films* **325**, 140 (1998).

46. B. A. Kolesov, T. V. Basova and I. K. Igumenov, *Thin Solid Films* **304**, 166 (1997).

47. R. Aroca, C. Jennings, R. O. Loutfy and A. M. Hor, *The Journal of Physical Chemistry* **90**, 5255 (1986).

48. J. Dowdy, J. J. Hoagland and K. W. Hipps, *The Journal of Physical Chemistry* **95**, 3751 (1991).

49. P. Y. Yu and M. Cardona, *Fundamentals of Semiconductors, Physics and Material Properties, 4th Edition* (Springer, Berlin Heidelberg, 2010.

50. E. Speiser, B. Buick, S. Del Gobbo and W. Richter, *Epioptics-9, Proceedings of the 38th Course of the International School of Solid State Physics, Erice, Italy, (World Scientific, Singapore, 2008)* (Ed. A. Cricenti, 82.

51. C. Jennings, R. Aroca, A.-M. Hor and R. O. Loutfy, *Spectrochimica Acta Part A: Molecular Spectroscopy* **41**, 1095 (1985).

52. L. Riele and T. Bruhn and V. Rackwitz and R. Passmann and B.-O. Fimland and N. Esser and P. Vogt, *Phys. Rev. B*, **84**, 205317 (2011).

53. L. Edwards and M. Gouterman, *Journal of Molecular Spectroscopy* **33**, 292 (1970).

# OPTICAL PROPERTIES OF NANOSTRUCTURED METAMATERIALS

Bernardo S. Mendoza*

*Division of Photonics, Centro de Investigaciones en Óptica, León, Guanajuato, México, * E-mail:bms@cio.mx*

W. Luis Mochán

*Instituto de Ciencias Físicas, Universidad Nacional Autónoma de México, A.P. 48-3, 62251 Cuernavaca, Morelos, México*

Guillermo Ortiz

*Departamento de Física, Facultad de Ciencias Exactas, Naturales y Agrimensura, Universidad Nacional del Nordeste - Instituto de Modelado e Innovación Tecnológica, CONICET-UNNE, Av. Libertad 5500, W3404AAS Corrientes, Argentina.*

Ernesto Cortés

*División de Ciencias e Ingenierías, Campus León, Universidad de Guanajuato, México*

We present a very efficient recursive method to calculate the effective optical response of nano-structured metamaterials made up of particles with arbitrarily shaped cross sections arranged in periodic two or three-dimensional arrays. We consider dielectric particles embedded in a metal matrix with a lattice constant much smaller than the wavelength of the incident field. Neglecting retardation our formalism allows factoring the geometrical properties from the properties of the materials. If the conducting phase is continuous the low frequency behavior is metallic. If the conducting paths are nearly blocked by the dielectric particles, the high frequency behavior is dielectric. Thus, extraordinary-reflectance and transparent bands may develop at intermediate frequencies, whose properties may be tuned by adjusting the geometry.

*Keywords*: Metamaterials, nano-structures, optical properties.

## 1. Introduction

Metamaterials are typically binary composites of conventional materials: a matrix with inclusions of a given shape, arranged in a periodic structure. Since the times of Maxwell, Lord Rayleigh and Maxwell-Garnet up to today, many authors have contributed to the calculation of the bulk macroscopic response in terms of the dielectric properties of its constituents.[1-3] Recent technologies allow the manufacture of ordered composite materials with periodic structures. For instance, high resolution electron beam lithography and its interferometric counterpart have been used in order to make particular designs of nano-structured composites, producing various shapes with nanometric sizes.[4,5] Moreover, ion milling techniques are capable of producing high quality air hole periodic and non-periodic two-dimensional (2D) arrays, where the holes can have different geometrical shapes.[6,7] Therefore, it is possible to build devices with novel macroscopic optical properties.[8] For example, a negative refractive index has been predicted and observed[9] for a periodic composite structure of a dielectric matrix with noble metal inclusions of trapezoidal shape.[10]

Nano-structured metallic films are having an important development as well. Metallic films with sub-wavelength nanometric holes may display an extraordinarily large transmittance at near infrared frequencies for which the metal is opaque and light waves are not expected to propagate within the holes.[11,12] The existence of surface plasmon-polariton (SPP) modes, may explain the enhancement of optical transmission through sub-wavelength holes.[13-16] Beside the single coupling to SPP modes, double resonant conditions[17] and waveguide modes[18] seem to play an important role in the optical enhancement for metallic gratings with very narrow slits and for compound gratings.[19] Also, a very strong polarization dependence in the optical response of periodic arrays of oriented sub-wavelength holes on metal hosts[6,7,20] and single rectangular inclusion within a perfect conductor[21] have been recently reported. These studies did not rely on SPP excitation as a mechanism to explain their optical results. The anomalous transmission and other extraordinary optical properties of nano-structured metallic films open the possibility of tailored design for many applications that include hyperlens far-field-subdiffraction imaging,[22-24] cloaking,[25] optical antennas,[26,27] and circular polarizers.[28] Thus, understanding the electromagnetic properties of this sometimes called plasmonic metamaterials has become important.

Many different approaches have been proposed to describe the extraordinary optical properties of some structured systems.[13-18,29-39] Some resonances have been identified with surface-plasmon-polaritons (SPP's) that may be excited by light after being scattered by the system.

In this work we obtain the macroscopic dielectric response of a periodic composite, using a homogenization procedure first proposed by Mochán and Barrera.[40] In this procedure the macroscopic response of the system is obtained from its microscopic constitutive equations by eliminating the spatial fluctuations of the field with the use of Maxwell's equations. Besides the average dielectric function, the formalism above incorporates the effects that the rapidly varying Fourier components of the microscopic response has on the macroscopic response, i.e. the local-field effect. Similar homogenization procedures are also found in.[41–44] In a recent work[20] we obtained the frequency-dependent complex macroscopic dielectric-response tensor $\epsilon_{ij}^M(\omega)$ of 2D-periodic lattices of cylindrical inclusions with arbitrarily shaped cross sections embedded within metallic hosts. We obtain that in the local limit an enhanced transparency is present without invoking explicitly a SPP mechanism.

In this review we develop Haydock's recursive scheme[45] to obtain the macroscopic dielectric response of 2D and 3D periodic metamaterials in the long wavelength limit. With this procedure one gains not only a tremendous speed improvement of several orders of magnitude over that of Ref. 20, but also the possibility of calculating the optical properties of sub-wavelength two-dimensional[46] and three-dimensional[47] structures with rather arbitrary geometry, including interpenetrated inclusions made out of dispersive and dissipative components. We show that the geometry of the inclusions and of the lattice might lead to very anisotropic optical behavior and to a very generic enhanced transmittance for metal-dielectric metamaterials whenever there are only poor conducting paths across the whole sample, or equivalently, that the transparency windows within metal-dielectric metamaterials appear for inclusion filling fractions slightly below the percolation threshold of the metallic phase.

## 2. Theory

We consider a metamaterial made of an homogeneous host of some material $a$ within which a periodic lattice of arbitrarily shaped nanometric inclusions of a material $b$ is embedded, yielding an artificial crystal. We assume that each region $\alpha = a, b$ is large enough though to have a well defined macroscopic response $\epsilon_\alpha$ which we assume local and isotropic. The lattice parameter is taken to be smaller than the free wavelength $\lambda_0 = 2\pi c/\omega$, with $c$ the speed of light in vacuum and $\omega$ the frequency. The microscopic response is described by

$$\epsilon(\mathbf{r}) = \epsilon_a - B(\mathbf{r})\epsilon_{ab} \tag{1}$$

where $\epsilon_{ab} \equiv \epsilon_a - \epsilon_b$ and $B(\mathbf{r}) = B(\mathbf{r} + \mathbf{R})$ is the periodic characteristic function for the $b$ regions, with $\{\mathbf{R}\}$ the Bravais lattice of the metamaterial.

The constitutive equation $\mathbf{D}(\mathbf{r}) = \epsilon(\mathbf{r})\mathbf{E}(\mathbf{r})$ may be written in reciprocal space as

$$\mathbf{D_G}(\mathbf{q}) = \sum_{\mathbf{G}'} \epsilon_{\mathbf{GG}'} \mathbf{E_{G}'}(\mathbf{q}), \tag{2}$$

where $\mathbf{D}(\mathbf{r})$ and $\mathbf{E}(\mathbf{r})$ are the electric and displacement fields, $\mathbf{D_G}(\mathbf{q})$ and $\mathbf{E_G}(\mathbf{q})$ the corresponding Fourier coefficients with wavevectors $\mathbf{q} + \mathbf{G}$, $\mathbf{q}$ is Bloch's vector and $\{\mathbf{G}\}$ the reciprocal lattice. Here, $\epsilon_{\mathbf{GG}'}$ is the Fourier coefficient of $\epsilon(\mathbf{r})$ corresponding to the wavevector $\mathbf{G} - \mathbf{G}'$. Ignoring retardation we may assume $\mathbf{E}$ is longitudinal,

$$\mathbf{E_G} \rightarrow \mathbf{E_G^L} = \hat{\mathbf{G}}\hat{\mathbf{G}} \cdot \mathbf{E_G}, \tag{3}$$

where we denote the unit vectors $(\mathbf{q}+\mathbf{G})/|\mathbf{q}+\mathbf{G}|$ simply by $\hat{\mathbf{G}}$, in particular, $\hat{\mathbf{0}} = \mathbf{q}/q$. A longitudinal external field may be identified with $\mathbf{D}^L$, which allows us to chose $\mathbf{D_{G \neq 0}^L}(\mathbf{q}) = 0$, i.e. we consider an external longitudinal plane wave without small scale spatial fluctuations. Substituting Eq. 3 into the longitudinal projection of Eq. 2 allows us to solve for

$$\mathbf{E_0^L} = \hat{\mathbf{q}}\eta_{00}^{-1}\hat{\mathbf{q}} \cdot \mathbf{D_0^L}, \tag{4}$$

where we first invert

$$\eta_{\mathbf{GG}'} \equiv \hat{\mathbf{G}} \cdot (\epsilon_{\mathbf{GG}'}\hat{\mathbf{G}}'). \tag{5}$$

and afterwards take the $\mathbf{00}$ component. The macroscopic longitudinal field $\mathbf{E}_{ML}$ is obtained from $\mathbf{E}^L$ by eliminating its spatial fluctuations, i.e., $\mathbf{E}_{ML} = \mathbf{E_0^L}$. Similarly, $\mathbf{D}_{ML} = \mathbf{D_0^L}$. Thus, from Eq. 4 we identify

$$\epsilon_{ML}^{-1} \equiv \hat{\mathbf{q}}\xi\hat{\mathbf{q}} = \hat{\mathbf{q}}\eta_{00}^{-1}\hat{\mathbf{q}}, \tag{6}$$

defined through $\mathbf{E}_{ML} = \epsilon_{ML}^{-1} \cdot \mathbf{D}_{ML}$, as the longitudinal projection of the macroscopic dielectric response corresponding to Bloch's wavevector $\mathbf{q}$,

To continue, Fourier transform the microscopic response, $\epsilon_{\mathbf{GG}'} = \epsilon_a\delta_{\mathbf{GG}'} - \epsilon_{ab}B_{\mathbf{GG}'}$, where

$$B_{\mathbf{GG}'} = (1/\Omega) \int_v d^3r\, e^{i(\mathbf{G}-\mathbf{G}')\cdot\mathbf{r}}, \tag{7}$$

$\Omega$ is the volume of the unit cell and $v$ the volume occupied by $b$. The geometry is characterized by $B_{\mathbf{GG}'}$ and in particular, $B_{00} = v/\Omega \equiv f$ is the filling fraction of the inclusions.

## 2.1. *Haydock's recursion*

From Eq. 5 we obtain $\eta_{\mathbf{GG'}}^{-1} = \mathcal{G}_{\mathbf{GG'}}/\epsilon_{ab}$, where $\hat{\mathcal{G}}(u) = (u - \hat{\mathcal{H}})^{-1}$ is a Green's function corresponding to an operator $\hat{\mathcal{H}}$ with elements

$$\mathcal{H}_{\mathbf{GG'}} = B_{\mathbf{GG'}}^{LL} = \hat{\mathbf{G}} \cdot (B_{\mathbf{GG'}}\mathbf{G'}), \qquad (8)$$

and where the frequency dependent spectral variable $u \equiv (1 - \epsilon_b/\epsilon_a)^{-1}$ is analogous to a complex energy.

From Eq. 6 we obtain $\xi = \langle 0|\hat{\mathcal{G}}(u)|0\rangle/\epsilon_{ab}$, where $|\mathbf{G}\rangle$ denotes a plane wave state with wave vector $\mathbf{q} + \mathbf{G}$. This allows the use of Haydock's recursive scheme to obtain the projected Green's function and thus the macroscopic response. We set $|-1\rangle = 0$, $|0\rangle = |0\rangle$, $b_0 = 0$ and recursively define the orthonormalized states $|n\rangle$ through

$$|\tilde{n}\rangle = \hat{\mathcal{H}}|n-1\rangle = b_{n-1}|n-2\rangle + a_{n-1}|n-1\rangle + b_n|n\rangle, \qquad (9)$$

with

$$a_{n-1} = \langle n-1|\tilde{n}\rangle = \langle n-1|\hat{\mathcal{H}}|n-1\rangle, \qquad (10)$$

and

$$b_n^2 = \langle \tilde{n}|\tilde{n}\rangle - a_{n-1}^2 - b_{n-1}^2. \qquad (11)$$

In the basis $\{|n\rangle\}$ the operator $\mathcal{H}$ may be represented by a tridiagonal matrix and the inverse $\mathcal{G}^{-1}(u) \equiv \mathcal{G}_0^{-1}(u)$ of the Green's function is given by the matrix

$$\mathcal{G}_0^{-1}(u) = \begin{pmatrix} u - a_0 & -b_1 & 0 & 0 & \cdots \\ -b_1 & u - a_1 & -b_2 & 0 & \cdots \\ 0 & -b_2 & u - a_2 & -b_3 & \cdots \\ 0 & 0 & -b_3 & u - a_3 & \ddots \\ \vdots & \vdots & & \ddots & \ddots \end{pmatrix}, \qquad (12)$$

which we can write recursively in blocks as

$$\mathcal{G}_n^{-1} = \left( \begin{array}{c|c} A_n & \mathcal{B}_{n+1} \\ \hline \mathcal{B}_{n+1}^T & \mathcal{G}_{n+1}^{-1} \end{array} \right), \qquad (13)$$

with $A_n = (u - a_n)$ and $\mathcal{B}_n = (-b_n, 0, 0, \cdots)$. Here we used calligraphic letters to denote any matrix except $1 \times 1$ matrices which are equivalent to scalars. Now we write $\mathcal{G}_n$ in blocks as

$$\mathcal{G}_n = \left( \begin{array}{c|c} R_n & \mathcal{P}_n \\ \hline \mathcal{Q}_n & \mathcal{S}_n \end{array} \right), \qquad (14)$$

so using $\mathcal{G}_n\mathcal{G}_n^{-1} = \mathrm{diag}(1)$ we find

$$R_n = \frac{1}{A_n - \mathcal{B}_{n+1}\mathcal{G}_{n+1}^{-1}\mathcal{B}_{n+1}^T} = \frac{1}{A_n - b_{n+1}^2 R_{n+1}}, \tag{15}$$

where in the last step we used the fact that the vectors $\mathcal{B}_{n+1}$ have only one element different from zero. In this way, we see the $n$-th solution is linked to the $n + 1$ solution. Iterating Eq. (15) we obtain $\mathcal{G}_{00}(u) = R_0$ and then

$$\xi = \frac{u}{\epsilon_a} \cfrac{1}{u - a_0 - \cfrac{b_1^2}{u - a_1 - \cfrac{b_2^2}{u - a_2 - \cfrac{b_3^2}{\ddots}}}}, \tag{16}$$

Notice that Haydock's coefficients depend only on the geometry through $B_{\mathbf{GG'}}^{LL}$. The dependence on composition and frequency is completely encoded in the complex valued spectral variable $u$. Thus, for a given geometry we may explore manifold compositions and frequencies without having to recalculate Haydock's coefficients.

We should emphasize that $\xi$ depends in general on the direction of $\mathbf{q}$. Calculating $\epsilon_{ML}^{-1}$ for several propagation directions $\hat{\mathbf{q}}$ we may obtain all the components of the full inverse long-wavelength dielectric tensor $\epsilon_M^{-1}$ and from it $\epsilon_M$.

To initiate the recursion in order to obtain $a_n$ and $b_n$ we first define the following auxiliary function

$$\varphi_n(\mathbf{G}) \equiv \langle\mathbf{G}|n\rangle. \tag{17}$$

Now we project Eq. 9 into $|\mathbf{G}\rangle$ and obtain

$$\varphi_{\tilde{n}}(\mathbf{G}) = \langle\mathbf{G}|\tilde{n}\rangle = \sum_{\mathbf{G'}} \mathbf{G} \cdot (B_{\mathbf{GG'}}\mathbf{G'})\varphi_{n-1}(\mathbf{G'}). \tag{18}$$

Using $\sum_{\mathbf{g}} |\mathbf{G}\rangle\langle\mathbf{G}| = 1$ and Eq. 10 we obtain

$$a_n = \langle n|\tilde{n}\rangle = \sum_{\mathbf{G}} \langle n|\mathbf{G}\rangle\langle\mathbf{G}|\tilde{n}\rangle = \sum_{\mathbf{G}} \varphi_n^*(\mathbf{G})\varphi_{\tilde{n}}(\mathbf{G}), \tag{19}$$

and $\langle\tilde{n}|\tilde{n}\rangle = \sum_{\mathbf{G}} |\varphi_{\tilde{n}}(\mathbf{G})|^2$, that when substituted in Eq. 11 gives $b_n$. Then from Eq. 9 we obtain

$$\varphi_n(\mathbf{G}) = \frac{\varphi_{\tilde{n}-1}(\mathbf{G}) - a_{n-1}\varphi_{n-1}(\mathbf{G}) - b_{n-1}\varphi_{n-2}(\mathbf{G})}{b_n}. \tag{20}$$

Employing above equations we can recursively calculate Haydock's coefficients $a_n$ y $b_n$ starting from $\varphi_{n-1}(\mathbf{G})$, besides obtaining $\varphi_n(\mathbf{G})$ with which we can start the next iteration till convergence is reached. We chose as the initial state $\varphi_0(\mathbf{G}) = \delta_{\mathbf{G0}}$ since the macroscopic dielectric function is given by the $\mathbf{G} = \mathbf{0}$, $\mathbf{G}' = \mathbf{0}$ component of Green's function.

We remark that since $B_{\mathbf{GG}'} = B(\mathbf{G} - \mathbf{G}')$, Eq. 18 is a convolution which according to Faltung's theorem may be obtained as the product of the characteristic function $B(\mathbf{r})$ with the inverse Fourier transform of $\hat{\mathbf{G}}'\varphi_{n-1}(\mathbf{G}')$ (denoted as $\vec{\varphi}_{n-1}(\mathbf{r})$). This result is of great numerical importance: by switching back and forth between real and reciprocal space we may obtain successive Haydock coefficients $a_n$ and $b_n$ through simple multiplications, without performing any large matrix products. We can perform calculations for an arbitrarily shaped inclusions simply by choosing the corresponding function $B(\mathbf{r})$ in real space.

Finally, a fast scheme to compute the continued fraction of Eq. 16 follows from the product,

$$\begin{pmatrix} p_n & p_{n-1} \\ q_n & q_{n-1} \end{pmatrix} \equiv \begin{pmatrix} u - a_o & 1 \\ 1 & 0 \end{pmatrix} \begin{pmatrix} u - a_1 & 1 \\ -b_1^2 & 0 \end{pmatrix} \cdots \begin{pmatrix} u - a_n & 1 \\ -b_n^2 & 0 \end{pmatrix}, \qquad (21)$$

from which we obtain

$$\xi = \frac{\epsilon_a}{u} \lim_{n \to \infty} \frac{p_n}{q_n}, \qquad (22)$$

where in practice a large but finite $n$ is needed to achieve convergence of the limit.

## 3. Results

We first compare our results to the previous formalism of Ortiz et al.[20] where a homogenization of Maxwell's equations was done without neglecting retardation. The retarded results do depend on the relative size of the unit cell and the wavelength of the incoming light. In Fig. 1 we show the calculated normal incidence reflectivity $R$[48] from a semi-infinite system made of an isotropic two-dimensional (2D) square array of cylindrical inclusions with $\epsilon_b = 4$ on a gold host (thus $\epsilon_a = \epsilon(\mathrm{Au})$ that is taken from Ref. 49) with filling fraction $f = 0.7$. When $\lambda_0 \sim L$ with $L$ the size of the unit cell, $R$ as obtained in Ref. 20 disagrees with our current calculation. This is not surprising, as here we have neglected retardation. However, for $\lambda_0 \gg L$ the two approaches agree within the numerical accuracy,[20] as could be expected. Also, we notice that $R$ for the metamaterial is rather different from that of pure gold. Indeed, we see that $R$ for the metamaterial is rather low at low frequencies and becomes almost zero at frequencies where gold is opaque and strongly reflective. We remark that for the chosen filling fraction, cylinders on different unit cells almost touch each other,

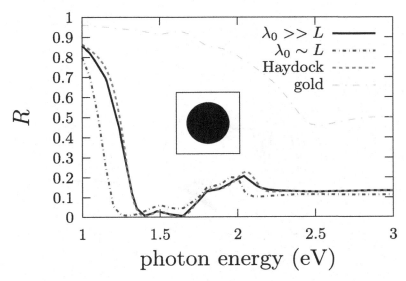

Fig. 1. Reflectance $R$ vs. photon energy for an isotropic 2D square array of cylindrical inclusion (see inset) with $\epsilon_b = 4$ and $f = 0.7$ on a gold host. $R$ of gold is shown for comparison, see text for details.

nearly choking the conducting paths. Thus, the system is dielectric like except at very small frequencies, where any small conductance dominates the macroscopic response. At intermediate frequencies the response of the metamaterial matches the dielectric constant of vacuum. This behaviour originates from the local-field effect and is determined by the geometry of the metamaterial.

We remark that following Ref. 20 requires the solution of a very large system of equations which took about 3 hours of CPU time spectra using 56 processors in parallel for each of the about 300 energy points we calculated for each spectrum in Fig. 1. In contrast, the calculation of Haydock's coefficients made on the interpreted Perl Data Language (PDL) took about 3 minutes of a single processor, and that they allow the immediate calculation of the whole spectra shown as well as any other spectrum for any other choice of materials. Thus, Haydock's method makes a huge difference in computing time.

In Figs. 2 we show $R_i$ $(i = x, y)$ for a 2D square array of 5-points stars, with $\epsilon_b = 4$. The results are converged by using $\sim 200$ $a_n$ and $b_n$ coefficients, and a real space grid of $\sim 400 \times 400$ points for $B(\mathbf{r})$. We see that for low $f$, $R_i$ is rather similar to that of gold, as one would expect. As $f$ grows towards

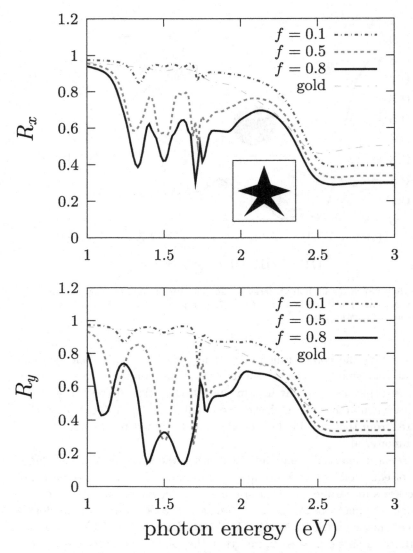

Fig. 2. $R_{x,y}$ vs the photon energy for an square 2D array of 5-points star inclusion (see inset) with $\epsilon_b = 4$ and various values of $f$ on a gold host. $R$ of gold is shown for comparison, see text for details.

the percolation threshold, we notice well defined low-energy minima where $R_i$ deviates from the metallic behavior. As in Fig. 1, their explanation is found in the change of behavior, from conducting at low-frequency to

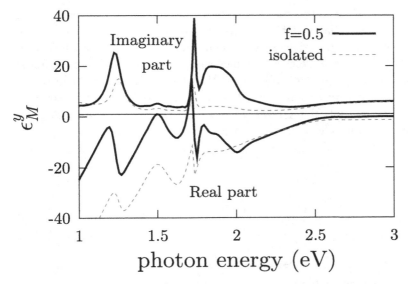

Fig. 3. $\epsilon_M^y$ vs the photon energy of the 5-points star system of Fig. 2 for $f = 0.5$ and an isolated 5-points star. The horizontal line is at one on the vertical scale.

dielectric at high frequencies. We see that the optical response is highly anisotropic, i.e. $R_x \neq R_y$, since this inclusion is geometrically anisotropic. The non-trivial behavior of $R$ occurs at infrared frequencies for which one would naively expect very high values for $R$. This anomalous reflection is due to excitation of resonances due to particular shape of the inclusion in the periodic array, and as for the case of the cylinders of Fig. 1, it is more apparent as $f$ increases towards the percolation threshold.[50]

In Fig. 3 we show the real and imaginary parts of the $y$ component of the macroscopic dielectric function $\epsilon_M^y$ for the 5-points star of Fig. 2 and an isolated 5-points star. First we notice that for $f = 0.5$ when $\text{Re}[\epsilon_M^y] = 1$ at 1.5 eV and 1.7 eV where $\text{Im}[\epsilon_M^y]$ is small, $R_y$ is close to zero as seen in Fig. 2, as one should expect since the macroscopic dielectric function is almost that of vacuum. However at 1.72 eV where again $\text{Re}[\epsilon_M^y] = 1$, but now $\text{Im}[\epsilon_M^y]$ is not small, $R_y$ is close to one. Also, we can see that the $\text{Im}[\epsilon_M^y]$ shows high absorption peaks (resonances) where regardless of the value of $\text{Re}[\epsilon_M^y]$, $R_y$ is close to one. We see that the line shape of $\epsilon_M^y$ is similar to that for $f = 0.5$, however the imaginary part is much smaller, meaning less absorption, and more importantly, the real part is never close to one. This in turn explains why $R_y$ for $f = 0.1$ is very close to that of pure gold. In a sense, an isolated inclusion is similar to a system with low filling fraction, since in this case the inclusions would be far from each other, like if they were isolated. Of course, above analysis could be done for any direction of $\epsilon_M$ and any given system. Thus one can see that the interaction through the local-field effect as the inclusions are closer together enhances the resonances seen in the $\text{Im}[\epsilon_M]$ and changes the $\text{Re}[\epsilon_M]$ in such a way that $R$ shows a very rich spectral dependence. Also, as we move towards the percolation limit, $\text{Re}[\epsilon_M]$ approaches and crosses one, thus giving the high-transmittance effect.

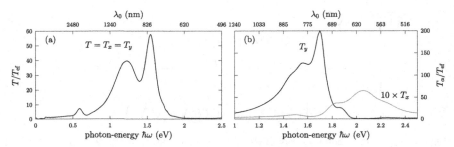

Fig. 4. Normal-incidence transmittance $T_\alpha$ for $\alpha = x, y$ polarization vs. frequency $\omega$ for 200 nm Au films with faces normal to the $z$ axis with an embedded lattice of dielectric inclusions, normalized to the transmittance $T_{\text{eff}}$ of a homogeneous Au film with the same amount of metal. (a) Simple cubic lattice of spheres of radius $r = 0.6a$ with $a$ the lattice parameter with $\epsilon_b = 4$. (b) Simple orthorhombic lattice of $z$-oriented cylinders with radius $r = 0.53a_x$, height $h = 0.9a_z$ and dielectric response $\epsilon_b = 2$ with lattice parameters $a_x = a_z$ and $a_y = 1.15a_x$.

In Fig. 4a we show the transmittance $T$ of a film made of a simple cubic lattice of spherical dielectric inclusions with response $\epsilon_b = 4$ within an Au host.[51] We chose the radius as $r = 0.6a$, with $a$ the lattice parameter, so the spheres overlap their neighbors. We have normalized the results to the transmittance $T_{\text{eff}}$ of an effective homogeneous Au film of width $d_{\text{eff}}$, in order to emphasize the transmittance enhancement due to the metamaterial geometry. Several enhancement peaks between one and two orders of magnitude are visible in the transmittance spectrum, corresponding to the excitation of coupled multipolar plasmon resonances within the region where the metal is opaque.

In Fig. 4b we show the normalized transmittance $T_\alpha/T_{\text{eff}}$ ($\alpha = x, y$) for plane polarized light normally incident on a 200 nm film lying on the $xy$ plane made of a simple orthorhombic lattice of $z$-oriented dielectric cylinders with radius $r = 0.53a_x$, height $h = 0.9a_z$ dielectric function $\epsilon_b = 2$ and lattice parameters $a_y = 1.15a_x$ and $a_z = a_x$ within an Au host. There is a huge anisotropy, with a peak enhancement of $T_y$ almost two orders of magnitude larger than that of $T_x$. For this geometry there is an overlap between neighbor cylinders along $x$, so the system is a better low frequency conductor along $y$.

## 4. Conclusions

We have developed a systematic scheme to calculate, the complex frequency dependent macroscopic dielectric function $\epsilon_M$ of metamaterials in terms of the dielectric functions of the host $\epsilon_a$ and the inclusions $\epsilon_b$, and of the geometry of both the unit cell and the inclusions. Starting from Maxwell's equations and employing a long wavelength approximation we have implemented the calculation through Haydock's recursive method which requires rather minimal computing resources to obtain well converged results. Our formalism may be employed to explore and design with a tailored optical response. In particular, we showed that extraordinary transparency of metamaterials is a rather generic phenomena whenever the conducting phase percolates and the metal surrounded inclusions display dielectric resonances. We hope this work motivates the experimental verification of our results through the construction and optical characterization of these systems.

## Acknowledgments

We acknowledge partial support from DGAPA-UNAM IN120909 (WLM), and FONCyT PAE-22592/2004 nodo NEA:23016 and nodo CAC:23831 (GPO).

## References

1. J.C.Garland and D.B.Tanner (eds.), *Electrical Transport and Optical Properties of Inhomogeneous Media*AIP Conference Proceeding No. 40, (American Institute of Physics, New York, 1978).
2. W. Mochán and R.G.Barrera (eds.), *Electrical Transport and Optical Properties of Inhomogeneous Media*Physica A 207, Num 1-3, (Elsevier, The Netherlands, 1994).
3. G. Milton, K. Golden, D. Dobson and A. Vardeny (eds.), *Electrical Transport and Optical Properties of Inhomogeneous Media*Physica B 338, Num 1-4, (Elsevier, North-Holland, 2003).
4. Y. Akahane, T. Asano, B.-S. Song and S. Noda, *Nature* **425**, 944 (2003).
5. A. N. Grigorenko, A. K. Geim, H. F. Gleeson, Y. Zhang, A. A. Firsov, I. Y. Khrushchev and J. Petrovic, *Nature* **438**, 335 (2005).
6. K. K. Koerkamp, S. Enoch, F. B. Segerink, N. van Hulst and L. Kuipers, *Phys. Rev. Lett.* **92**, p. 183901 (2004).

7. R. Gordon, A. G. Brolo, A. McKinnon, A. Rajora, B. Leathem and K. L. Kavanagh, *Phys. Rev. Lett.* **92**, p. 037401 (2004).
8. J. Pendry, *Phys. Rev. Lett.* **85**, p. 3966 (2000).
9. V.M.Shalaev, W. Cat, U. Chettiar, H. Yuan, A. Sarychev, V. Drachev and A. Kildishev, *Opt. Lett.* **30**, 3356 (2005).
10. A. Kildishev, W. Cai, U. Chettiar, . H.-K. Yuan, A. Sarychev, V. P. Drachev and V. M. Shalaev, *J. Opt. Soc. Am. B* **23**, 423 (2006).
11. L. Novotny and B. Hecht, *Principles of Nano-Optics* (Cambridge University Press, New York, 2006).
12. J. Weiner, **72**, p. 0644011 (2009).
13. P. Sheng, R. Stepleman and P. Sanda, *Phys. Rev. B* **26**, p. 2907 (1982).
14. H. Lochbihler and R. Depine, *Appl. Opt.* **32**, p. 3459 (1993).
15. H. Lochbihler, *Phys. Rev. B* **50**, p. 4795 (1994).
16. H. Ghaemi, T. Thio, D. Grupp, T. Ebbesen and H. Lezec, *Phys. Rev. B* **58**, p. 6779 (1998).
17. S. Darmanyan and A. Zayats, *Phys. Rev. B* **67**, p. 035424 (2003).
18. J. Porto, F. García-Vidal and J. Pendry, *Phys. Rev. Lett.* **83**, p. 2845 (1999).
19. D. Skigin and R. Depine, *Phys. Rev. Lett.* **95**, 217402 (2005).
20. G. P. Ortiz, B. E. Martínez-Zérega, B. S. Mendoza and W. L. Mochán, *Phys. Rev. B* **109**, p. 245132 (2009).
21. F. García-Vidal, E. Moreno, J. Porto and L. Martín-Moreno, *Phys. Rev. Lett.* **95**, p. 103901 (2005).
22. L. Chen and G. P. Wang, *Opt. Express* **17**, p. 3903 (2009).
23. Z. Liu, H. Lee, Y. Xiong, C. Sun and X. Zhang, *Science* **315**, p. 1686 (2007).
24. A. Salandrino and N. Engheta, *Phys. Rev. B* **74**, p. 075103 (2006).
25. W. Cai, U. K. Chettiar, A. V. Kildishev and V. M. Shalaev, *Appl. Phys. Lett* **91**, p. 1111051 (2007).
26. L. Novotny, *Phys. Rev. Lett.* **98**, p. 2668021 (2007).
27. W. Dickson, G. Wurtz, P. Evans, D.O'Connor, R. Atkinson, R. Pollard and A. Zayats, *Phys. Rev. B* **76**, p. 115411 (2007).
28. J. K. Gansel, M. Thiel, M. S. Rill, M. Decker, K. Bade, V. Saile, G. von Freymann, S. Linden and M. Wegener, *Science* **325**, p. 1513 (2009).
29. B.Hou, H.Wen, Y.Leng and W.Wen, **87**, p. 217 (2007).
30. L. Zhou, W. Wen, C.T.Chan and P. Sheng, *Phys. Rev. Lett.* **94**, p. 2439051 (2005).
31. J.B.Pendry, L.Martín-Moreno and F.J.Garcia-Vidal, *Science* **305**, p. 847 (2004).
32. L.Martín-Moreno, F.J.García-Vidal, H.J.Lezec, K.M.Pellerin, T.Thio, J.B.Pendry and T.W.Ebbesen, *Phys. Rev. Lett.* **86**, p. 1114 (2001).
33. D. Grupp, H. Lezec, T. Ebbesen, K. Pellerin and T. Thio, *Appl. Phys. Lett* **77**, p. 1569 (2000).
34. Q.-H. Park, J.H.Kang, J.W.Lee and D.S.Kim, *Opt. Express* **15**, p. 6994 (2007).
35. A. Agrawal, Z. Vardeny and A. Nahata, *Opt. Express* **16**, p. 9601 (2008).
36. Q. Cao and P. Lalanne, *Phys. Rev. Lett.* **88**, p. 0574031 (2002).
37. M. Treacy, *Appl. Phys. Lett* **75**, p. 606 (1999).

38. M. Treacy, *Phys. Rev. B* **66**, p. 1951051 (2002).
39. E.Popov, M.Nevière, S.Enoch and R.Reinisch, *Phys. Rev. B* **62**, p. 16100 (2000).
40. W.L.Mochán and R.G.Barrera, *Phys. Rev. B* **32**, p. 4984 (1985).
41. P. Halevi and F. Pérez-Rodríguez, *SPIE* **6320** (2006).
42. A.A.Krokhin, P. Halevi and J.Arriaga, *Phys. Rev. B* **65**, 115208 (2002).
43. P. Halevi, A.A.Krokhin and J.Arriaga, *Phys. Rev. Lett.* **82**, 719 (1999).
44. S. Datta, C. T. Chan, K. M. Ho and C. M. Soukoulis, *Phys. Rev. B* **48**, 14936 (1993).
45. R.Haydock, *Solid State Physics* **35**, p. 215 (1980).
46. E. Cortés, W. L. Mochán, B. S. Mendoza and G. P. Ortiz, *Phys. Status Solidi B* **247**, p. 2102 (2010).
47. W. L. Mochán, G. P. Ortiz and B. S. Mendoza, *Opt. Express* **18**, p. 22119 (2010).
48. The normal incidence reflectivity is given by the standard formula $R_i = (\sqrt{\epsilon_M^i} - 1)/(\sqrt{\epsilon_M^i} + 1)$, with $i$ a principal Cartesian direction.
49. P. Johnson and R. Christy, *Phys. Rev. B* **6**, p. 4370 (1972).
50. This limit is close to $f = 0.8$. The percolation limit we are referring to is that where the inclusion of a given unit cell just touches the inclusions of the neighbouring cells.
51. P. Johnson and R. Christy, *Phys. Rev. B* **6**, p. 4370 (1972).

# AB INITIO LONG-WAVELENGTH PROPERTIES OF METALLIC SYSTEMS: IRON AND MAGNESIUM

M. CAZZANIGA*[1-2], L. CARAMELLA[1-2], N. MANINI[1-2], P. SALVESTRINI[1-3]
and G. ONIDA[1-2]

[1] *Università degli Studi di Milano, Physics Department, Milan, Italy*
[2] *European Theoretical Spectroscopy Facility (ETSF)*
[3] *CNR-IFN Milan, Italy*
*E-mail: marco.cazzaniga@unimi.it*

In this work we describe a method, based on the linearization of the band dispersion, to compute the intraband contributions to the optical spectra of metals. To investigate its performance we test it on cubic iron and hexagonal magnesium. We compute the dielectric function and optical conductivity of these metals based on *ab initio* DFT-LDA band structure, in the RPA with a full inclusion of local-field effects. In particular we show that our method is capable to describe the anisotropic response of noncubic metals, without the need to resort to phenomenological parameterizations of the plasmon-pole type. We also introduce a method to recover the correct asymptotic trend in the $\omega \to 0$ limit and hence, the experimental value of the static conductivity.

*Keywords*: Optical properties; metals; TDDFT; Intraband contributions

## 1. Introduction

Due to screening, *ab initio* calculations of the optical properties of metallic systems are less problematic than those of insulating material from the point of view of electronic correlations, and indeed the random phase approximation (RPA) has been found to work fairly well. On the other hand, metals present an additional difficulty due to the need of an accurate description of the interband transitions. Standard approaches[1,2] rely on the addition of a Drude contribution to the intraband dielectric response. Such kind of approach has two obvious shortcomings: (i) it is intrinsically isotropic; (ii) it ignores the crystal local field (CLF). Despite for many systems these constrains constitute reasonable approximations a general treatment is desirable anyway.

In the present work[3] we propose a technical improvement aimed at overcoming the difficulties discussed above. We choose to expand the difference of the energies between the states involved in the intraband transitions by a first order expansion. This solution permits to avoid numerical instabilities which can arise if a brute force evaluation is performed. In addition the inclusion of the intraband transitions into the response function allows one to include CLF straightforwardly and to treat anisotropic systems.

After a brief review of the standard approach to calculate optical properties in Sec. 2, we present our method to deal with the Drude contribution in Sec. 3. This method is tested on magnetic bulk iron (Sec. 4.1) and weakly anisotropic bulk magnesium (Sec. 4.2). The numerical results for these metals are compared with available experimental results.[4-6]

## 2. Theoretical background

Calculations of optical properties are performed in time-dependent density functional theory (TDDFT), within linear-response.[7,8] In practice this approach requires the numerical evaluation of the independent-particle polarizability $\chi^{(0)}$:

$$\chi^{(0)}_{\sigma,\sigma',\mathbf{G},\mathbf{G}'}(\mathbf{q},\omega) = \frac{1}{V_{BZ}} \delta_{\sigma,\sigma'} \sum_{j,j'} \int_{\Omega} d^3k \frac{f(\epsilon_{j',\sigma}(\mathbf{k}+\mathbf{q})) - f(\epsilon_{j,\sigma}(\mathbf{k}))}{\omega^+ - [\epsilon_{j',\sigma}(\mathbf{k}+\mathbf{q}) - \epsilon_{j,\sigma}(\mathbf{k})]} \tag{1}$$

$$\times \langle \mathbf{k}, j, \sigma | e^{-i(\mathbf{q}+\mathbf{G})\cdot\hat{\mathbf{r}}} | \mathbf{k}+\mathbf{q}, j', \sigma \rangle \langle \mathbf{k}+\mathbf{q}, j', \sigma | e^{i(\mathbf{q}+\mathbf{G}')\cdot\hat{\mathbf{r}}} | \mathbf{k}, j, \sigma \rangle,$$

where the expression $\omega^+$ means that a small imaginary part is added to the frequency to account for the retarded response. The RPA macroscopic dielectric function is then obtained as:

$$\varepsilon_M(\mathbf{q},\omega) = \frac{1}{[\varepsilon^{-1}_{\mathbf{G},\mathbf{G}'}(\mathbf{q},\omega)]_{0,0}}, \tag{2}$$

where the microscopic dielectric function $\varepsilon$ is related to $\chi^{(0)}$ by:

$$\varepsilon_{\mathbf{G},\mathbf{G}'}(\mathbf{q},\omega) = \delta_{\mathbf{G},\mathbf{G}'} - \mathrm{Tr}_\sigma \left[ v_C(\mathbf{q}+\mathbf{G}) \chi^{(0)}_{\sigma,\sigma'\mathbf{G},\mathbf{G}'}(\mathbf{q},\omega) \right]. \tag{3}$$

To evaluate the interband contribution to the long wavelength limit of $\chi^{(0)}$ the standard approach is based on the expansion of the matrix element appearing in Eq. (1):

$$\langle \mathbf{k}, j, \sigma | e^{-i\mathbf{q}\cdot\hat{\mathbf{r}}} | \mathbf{k}+\mathbf{q}, j', \sigma \rangle \simeq \delta_{j,j'} + \frac{\mathbf{q} \cdot \langle \mathbf{k}, j, \sigma | i\nabla + i[V_{nl}, \mathbf{r}] | \mathbf{k}, j', \sigma \rangle}{\epsilon_{j',\sigma}(\mathbf{k}) - \epsilon_{j,\sigma}(\mathbf{k})}, \tag{4}$$

where the contribution coming from $[V_{nl}, \mathbf{r}]$ is relevant only if non-local pseudopotentials are used.

This procedure has been found to work well for insulating materials, while fails in the case of metals since the intraband transition cannot be described, due to the vanishing occupancy difference in Eq. (1).

## 3. The intraband contributions

To deal with the Drude contribution we employ a perturbative approach to estimate the vanishing difference of energies:

$$\Delta \epsilon_{j,\sigma}(\mathbf{k}) = \epsilon_{j,\sigma}\left(\mathbf{k} + \frac{1}{2}\mathbf{q}\right) - \epsilon_{j,\sigma}\left(\mathbf{k} - \frac{1}{2}\mathbf{q}\right) \simeq \mathbf{q} \cdot \nabla_{\mathbf{k}}\epsilon_{j,\sigma}(\mathbf{k}) . \quad (5)$$

The band energy gradient is evaluated as matrix element of the velocity operator:

$$\nabla_{\mathbf{k}}\epsilon_{j,\sigma}(\mathbf{k}) = \langle \mathbf{k}, j, \sigma | \frac{d\mathbf{r}}{dt} | \mathbf{k}, j, \sigma \rangle = \langle \mathbf{k}, j, \sigma | i[H, \mathbf{r}] | \mathbf{k}, j, \sigma \rangle \quad (6)$$

$$= \langle \mathbf{k}, j, \sigma | \mathbf{p} | \mathbf{k}, j, \sigma \rangle + \langle \mathbf{k}, j, \sigma | i[V_{\mathrm{nl}}, \mathbf{r}] | \mathbf{k}, j, \sigma \rangle .$$

The knowledge of $\Delta\epsilon_{j,\sigma}(\mathbf{k})$ allows us to evaluate the difference of the occupancies at the numerator of Eq. (1). Rather than expanding the highly nonlinear Fermi function, we prefer to consider the small-$\mathbf{q}$ expansion of its argument, and take the explicit difference:

$$\Delta f_{j,\sigma}(\mathbf{k}) \simeq f\left(\epsilon_{j,\sigma}(\mathbf{k}) + \frac{\Delta\epsilon_{j,\sigma}(\mathbf{k})}{2}\right) - f\left(\epsilon_{j,\sigma}(\mathbf{k}) - \frac{\Delta\epsilon_{j,\sigma}(\mathbf{k})}{2}\right). \quad (7)$$

The last ingredients are the matrix elements $\langle \mathbf{k} - \mathbf{q}/2, j, \sigma | e^{-i\mathbf{q}\cdot\hat{\mathbf{r}}} | \mathbf{k} + \mathbf{q}/2, j, \sigma \rangle$ which are retained at their zeroth order in $\mathbf{q}$, as in Eq. (4).

By symmetrizing the expression thanks to time-reversal symmetry, the intraband contribution to $\chi^{(0)}$ can be written:

$$\chi_{\mathbf{G},\mathbf{G}'}^{(0)\,\mathrm{intra}}(\mathbf{q}, \omega) = -\frac{1}{2\Omega} \sum_{j} \sum_{\sigma} \int_{\Omega} d^3k \, \Delta f_{j,\sigma}(\mathbf{k}) \quad (8)$$

$$\left[ \frac{1}{\omega^+ - \Delta\epsilon_{j,\sigma}(\mathbf{k})} - \frac{1}{\omega^+ + \Delta\epsilon_{j,\sigma}(\mathbf{k})} \right].$$

It is a known problem that the use of an expression like Eq. (8) with a constant imaginary part added to the frequency yields to an incorrect asymptotics for the $\omega \to 0$ behaviour of the imaginary part of the dielectric function. In particular such an expression would give a $\varepsilon_2 \propto \omega^{-3}$ (rather than the correct $\varepsilon_2 \propto \omega^{-1}$), yielding an unphysical diverging conductivity. To overcome this problem, we modify $\omega^+$ by introducing a suitable frequency dependency for $\omega^+$, in order to change the asymptotics for small frequencies while recovering the usual behaviour as soon as the frequency increases. Explicitly, we take

$$\omega^+ = \sqrt{\omega(\omega + i\eta)}, \tag{9}$$

with real $\omega$.

## 4. Applications

The procedure described in Sec. 3 has been implemented in the dp code.[9] Test calculation has been performed for bulk iron and magnesium at the respective experimental crystal structures. Ground states are computed with DFT-LDA, using the abinit package.[10,11] For both materials we choose a random sampling of the Brillouin zone with 40000 k-points (in the case of magnesium we need to symmetrize the mesh to be able to calculate the small anisotropy). The occupancies are smeared with a Fermi function based on a smearing temperature of $T_{\text{smear}} = 1 \cdot 10^{-5}$ Ha. CLF effects are taken into account by inverting the $\varepsilon$ matrix on a basis of 55 G-vectors for iron and 43 G-vectors for magnesium.

The choice of the q vectors is rather crucial: they should be tiny enough to avoid finite-q artifacts in the small-$\omega$ asymptotics. We select $\mathbf{q} = (5.80, 4.64, 3.48) \cdot 10^{-6} a_0^{-1}$ for iron, while to explore the anisotropies of magnesium, its response function has been computed for $\mathbf{q} = (1.03, -0.60, 0) \cdot 10^{-7} a_0^{-1}$ (in plane direction) and $\mathbf{q} = (0, 0, 0.63) \cdot 10^{-7} a_0^{-1}$ (c-axis).

$\varepsilon_2$ depends also on the $\eta$ appearing in Eq. (9). This parameter mimics the finite quasiparticle lifetime, mainly due to electrons scattering against phonons and/or crystal-defects, a sample-dependent quantity which is problematic to determine *ab initio*. In this work we estimate it based on a fit on the available experimental data in the infrared region,[4] consistent with $\eta = 4.60$ meV for Fe and $\eta = 140$ meV for Mg. To better display the spectral features at optical frequencies we use a different value of $\eta = 50$ meV for the interband contribution of both metals.

### 4.1. *Iron*

Figure 1 reports the computed dielectric function and the optical conductivity for bulk iron. By considering separately intra- and inter-band contributions of $\chi^{(0)}$ it is possible to observe that intraband response dominates at low frequency $\omega < 0.5$ eV, while interband transitions take over for optical frequencies. Fig. 1 shows that in addition the proposed approach restores the expected infrared power-law crossover from $\varepsilon_2 \simeq \omega_p^2/(\eta\,\omega)$, to $\varepsilon_2 \simeq \omega_p^2 \eta/\omega^3$ occurring near $\omega \simeq \eta$.

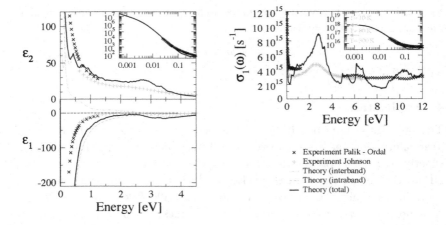

Fig. 1. Compared theoretical and measured dielectric function (left) and optical conductivity (right) for iron. Experiments from Refs. 4,5.

When compared with experiment, our results show overall agreement of the few discernible experimental features. In particular the absolute value of $\varepsilon$ and $\sigma$ and the general spectral shape are in fair agreement with the measured ones, especially below 8 eV. The intraband peak presents a correct overall scale due to the value of $\eta$ being fitted precisely to the near-infrared part of the experimental spectrum. The static limit of the conductivity turns out to be $1.4 \cdot 10^{18}$ s$^{-1}$, a value which falls in the range of measured low-temperature resistivities.[12] Discrepancies in the $8 - 12$ eV features might be due to lifetime and correlation effects not accounted for in the present simple model,[13] and which might be captured by a more refined theoretical approach, such as GW or Bethe-Salpeter calculation.

An analysis of the origin of the structures attributes the peaks around 2.8 eV and 6.5 eV as mainly due to the minority spin, while the majority spin dominates the spectrum in the region above 8 eV.

### 4.2. Magnesium

We compute the dielectric function as well as the optical conductivity for polarization perpendicular and parallel to the hexagonal crystal $c$ axis. Results are reported in Fig. 2. The calculations predict a 28% anisotropy in the spectral region near 0.7 eV, mainly associated to interband transitions. In that energy region computed spectra predict a stronger absorption in the basal plane than along the $c$-axis, in accord with single-crystal experimental

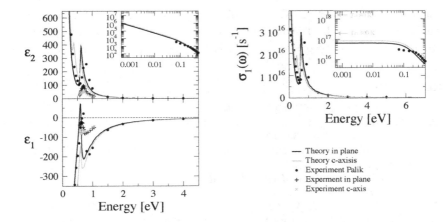

Fig. 2. Compared theoretical and measured dielectric function (left) and optical conductivity (right) for magnesium. Experiments from Refs. 4,6.

evidence.[6] On the contrary, the anisotropy in the far infrared region is opposite to the one just discussed. In detail the optical conductivity is larger for $c$-axis polarization than in the basal plane, by a significant, although moderate, 30%. In addition the static limit of $\sigma$ ($6.5 \cdot 10^{16}$ s$^{-1}$ in plane and $8.7 \cdot 10^{16}$ s$^{-1}$ $c$-axis), is slightly smaller than the measured room temperature resistivity of magnesium,[12] probably due to a overestimation of the parameter $\eta$ obtained from the fit of the experimental infrared response data.

## 5. Conclusions

In this work we present an *ab initio* approach to the calculation of the optical properties of metallic systems, by addressing the problems of intraband transitions and of the correct small frequency asymptotics. Tests on bulk iron and magnesium revealed the RPA dielectric function to be in significant accord with the experimentally observed spectral features. The calculations on magnesium have shown that our approach is capable to deal also with anisotropic systems.

## Acknowledgments

The authors acknowledge R. Del Sole, A. Marini, E. Mulazzi, L. Reining, F. Sottile for useful discussions. This work was funded in part by the EU's 7th Framework Programme through the European Theoretical Spectroscopy

Facility e–I3 (Grant No. INFRA-2007-211956). L.C. thanks the Università Italo-Francese for the financial support through the "Borsa Vinci 2006". We acknowledge generous supercomputing support from CINECA.

## References

1. K.-H. Lee and K. J. Chang, *Phys. Rev. B* **49**, 2362 (1994).
2. E. G. Maksimov, I. I. Mazin, S. N. Rashkeev and Y. A. Uspensky, *J. Phys. F: Met. Phys.* **18**, p. 833 (1988).
3. M. Cazzaniga, L. Caramella, N. Manini and G. Onida, *Phys. Rev. B* **82**, p. 035104 (2010).
4. E. D. Palik, *Handbook of optical constants of solids* (Academic Press, San Diego, 1998).
5. P. B. Johnson and R. W. Christy, *Phys. Rev.* **9**, p. 5056 (1974).
6. D. Jones and A. H. Lettington, *Proc. Phys. Soc.* **92**, p. 948 (1967).
7. E. Runge and E. K. U. Gross, *Phys. Rev. Lett.* **52**, p. 997 (1984).
8. S. Botti, A. Schindlmayr, R. Del Sole and L. Reining, *Rep. Progr. Phys.* **70**, p. 357 (2007).
9. http://www.dp-code.org.
10. http://www.abinit.org.
11. X. Gonze *et al.*, Computer Phys. Commun. **180**, p. 2582 (2009).
12. R. C. Weast and M. J. Astle, *CRC Handbook of Chemistry and Physics $63^{rd}$ edition* (CRC Press inc., Boca Raton, Florida USA, 1982).
13. A. Marini and R. Del Sole, *Phys. Rev. Lett.* **91**, p. 176402 (2003).

# MAGNETIC SECOND-HARMONIC GENERATION FROM INTERFACES AND NANOSTRUCTURES

J.F. MCGILP[†]

*School of Physics, Trinity College Dublin, Dublin 2, Ireland*

Magneto-optic techniques provide non-contact and non-destructive characterization of magnetic materials. This includes embedded magnetic nanostructures, which are accessible due to the large penetration depth of optical radiation. Nonlinear magnetic second-harmonic generation (MSHG) can measure the surface and interface magnetism of centrosymmetric magnetic films and nanostructures with sub-monolayer sensitivity. MSHG is briefly reviewed and examples from high symmetry interfaces and nanostructures described. Low symmetry systems, such as aligned magnetic nanostructures grown by self-organization on vicinal substrates, are more difficult to characterize, however, because of the large number of tensor components that may contribute to the signal. Normal incidence geometry simplifies the problem and this new approach is shown to allow the determination of hysteresis loops and the temperature dependent magnetic response of Au-capped Fe nanostructures grown on vicinal W(110).

## 1. Introduction

Magnetic interfaces and nanostructures are attracting interest in relation to their fundamental physics and technological application [1]. Self-assembly at atomic steps has been used to grow aligned, model nanostructures: for example, Fe nanostripes have been grown on vicinal W(110) and their magnetic behaviour has been probed using the magneto-optic Kerr effect (MOKE) [2, 3], while x-ray magnetic circular dichroism (XMCD) has been used to identify ferromagnetism in single atomic wires of Co grown on Pt(997) [4, 5]. A key advantage of these optical techniques, apart from their sensitivity, is their ability to probe buried nanostructures located in the interfacial region between the substrate and the capping layer used to protect the nanostructure from environmental corrosion and contamination [6]. However, techniques such as MOKE and XMCD have

---

[†] E-mail address: jmcgilp@tcd.ie

difficulty in discriminating between bulk and interface effects, or between interface and step contributions in aligned magnetic nanostructures.

Nonlinear optical techniques, such as optical second-harmonic generation (SHG), use symmetry to discriminate between the interfacial contribution and the normally dominant bulk response, allowing the interfacial structure of centrosymmetric systems to be determined [7]. Magnetic SHG (MSHG) extends this approach to magnetic interfaces. Magnetization, as an axial vector, does not lift the inversion symmetry of the bulk, allowing magnetic surfaces and interfaces to be probed [8, 9]. The first experimental results appeared in 1991 [10] and, with the development of reliable femtosecond lasers, MSHG surface and interface studies became relatively straightforward, due to the improvement in the signal-to-noise ratio (SNR).

Within the electric dipole approximation, the intensity of MSHG from a magnetic interface is given by:

$$I(2\omega, \pm M) \propto | \chi_{even}^{eff} E(\omega)E(\omega) \pm \chi_{odd}^{eff} M E(\omega)E(\omega) |^2 \qquad (1)$$

where $\chi_{even}^{eff}$ is the effective third rank crystallographic susceptibility tensor, $E(\omega)$ is the input electric field vector, $\chi_{odd}^{eff}$ is the effective fourth rank axial magnetic susceptibility tensor, and $M$ is the interface magnetization [9]. Higher order quadrupolar crystallographic contributions from the substrate or capping layer can be included in the effective value of the *even* term. SHG is known to be sensitive to strain [11] and any magneto-elastic contributions will appear in the *even* term [12]. Appropriate Fresnel and local electromagnetic field factors [13] are included in the effective tensor components of Eq. (1).

A comprehensive review of MSHG has been published recently by Kirilyuk and Rasing [14] and only a few important examples will be highlighted here. Compared to SHG, new *odd* magnetic tensor components have appeared, giving a contribution to the MSHG intensity that changes sign with the magnetization. The crystallographic terms are *even* in the magnetization, as shown in Eq. (1). This analysis was elegantly confirmed when an optical phase shift of 180° in the SH signal was measured on reversing the magnetization in a Rh/Co/Cu multilayer [15]. Hysteresis loops in the MSHG intensity have been measured for different magnetic surface and interfaces. The Cu(001)/Fe system shows 4x1 and 5x1 reconstructions below 4 monolayers (ML) coverage, and a 2x1 reconstruction above 4 ML. Very different surface hysteresis loops for these two phases were found (Figure 1), and coverage dependent studies up to 12 ML were used to show that there was no bulk contribution to the MSHG signal [16]. Other

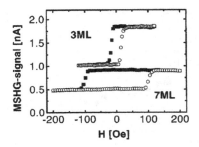

Figure 1. MSHG hysteresis loops for 3 ML and 7 ML Fe on Cu(001) (after [16]).

important experiments include the *in situ* measurement of oscillations in the MSHG intensity during the growth of Co and Fe films [17, 18]. In the latter case a monolayer of oxygen acts as a surfactant in the homoepitaxial growth of Fe films on Fe(001), floating on the top of the growing film. The oscillations in MSHG intensity (Figure 2) were attributed to the 7% outward relaxation of the

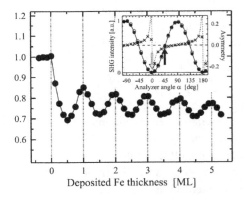

Figure 2. MSHG Normalized effective odd component amplitude for p-in/s-out geometry, as a function of the deposited Fe thickness on Fe(001)-1x1-O. Inset: SHG intensities, $I^+$(open circles) and $I^-$ (solid circles), and asymmetry, $A$ (crosses) as a function of the analyzer angle for p-in fundamental light. The p-in/s-out and p-in/sp-out geometries are indicated by the gray and black arrows (after [18]).

top Fe layer, which is expected to increase the magnetic moment of the Fe atoms [18]. *Ex situ* MSHG and MOKE oscillations have been observed from an exchange biased Co wedge, grown layer-by-layer on Cu(001) from 9 to 13 ML coverage, and capped with a 25 ML Mn film [19]. Coercivities determined from

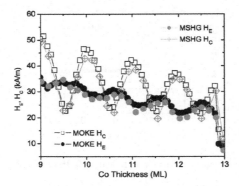

Figure 3. Coercivity and exchange bias of the hysteresis curves from MSHG (in gray) and MOKE (in black) as function of the Co thickness (after [19]).

the MSHG and MOKE measurements oscillate in phase (Figure 3), indicating that the bulk and interface magnetization change in the same way with field reversal. The MSHG asymmetry or contrast was used to infer that the net magnetic moment at the interface was enhanced at monolayer completion. Assuming that the effective *odd* contribution is relatively small compared to the *even* contribution, the expression for the MSHG contrast can be linearized:

$$A = \frac{I^+ - I^-}{I^+ + I^-} \approx 2\frac{|\chi_{odd}^{eff} M|}{|\chi_{even}^{eff}|}\cos\theta \qquad (2)$$

where the +/- superscript refers to equal but opposite magnetizations, and $\theta$ is the phase difference between the *odd* and *even* components [16, 18, 19]. It can be seen from Eq. (2) that, if the effective *even* and *odd* components either remain constant, or vary in the same way, in an experiment, the asymmetry is proportional to the magnetization. For example, the main contributions to variation in the tensor components with coverage are expected to come from changes in the local electronic structure and the local electromagnetic fields at the interface. It was pointed out that the edge or step contributions are likely to be different to those of the islands or terrace [19]. Indeed, MOKE studies from magnetic films grown on vicinal substrates have reported distinct step effects at higher vicinal angles [20-22].

## 2. Extension to aligned nanostructures

Step or edge contributions are expected to become increasingly important in the magnetization behaviour of aligned magnetic nanostructures as their dimensions shrink. Self-assembly at atomic steps on vicinal single crystal surfaces has proved to be a useful route to aligned structures, with two well studied systems being vicinal W(110)/Fe, where MOKE was used to measure the magnetization of sub-monolayer nanostripes as small as 10 atoms in width [2], and Pt(997)/Co, where the magnetization of single atomic wires was measured using XMCD [4]. A major attraction of applying MSHG to aligned magnetic nanostructures is that the symmetry of the edges or steps is lower than that of the terraces or islands and this, in principle, allows their contribution to the response of the system to be distinguished by MSHG [23]. However, two major difficulties must first be overcome: sensitivity and complexity.

One approach to overcoming the former is to account properly for the quadratic magnetic response implicit in Eq. (1), rather than choosing an experimental configuration that produces a smaller magnetic response in order to allow Eq. (1) to be linearized [24]. It may also be difficult to judge whether quadratic terms are distorting the loops in exchange biased systems, where loops may be acentric. Figure 4 shows a comparison of MOKE and SHG results from

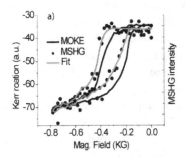

Figure 4. Coercivity and exchange bias of the hysteresis curves from MSHG (in gray) and MOKE (in black) as function of the Co thickness (after [25]).

an exchange biased system, where it was pointed out that the difference may be due to quadratic effects [25]. Without this constraint, a large value of the asymmetry can be chosen, which increases the SNR.

The difficulties arising from the complexity of the nonlinear response have resulted in MSHG studies being restricted to magnetic systems of high surface and interface symmetry [14]. Lower symmetry systems, which may have multiple magnetic regions, have many tensor components that may contribute to the MSHG intensity, making interpretation particularly difficult. Recently, a new approach to MSHG studies of lower symmetry systems has been developed and applied to the vicinal W(110)/Fe system, capped with Au [26, 27]: normal incidence (NI) SHG geometry simplifies the nonlinear response from systems of lower symmetry by excluding $z$-dependent tensor components. Fine tuning of the input polarization direction enhances the magnetic contribution to the NI SHG signal, resulting in a significant improvement in the SNR of hysteresis loops extracted from the measurements. The improved SNR allows the presence of different magnetic regions at the inhomogeneous interface to be identified by changing the input polarization direction, which alters relative contribution of the tensor components to the overall NI MSHG response.

## 3. NI MSHG from magnetic nanostructures of 1m symmetry

In general, low symmetry magnetic interfaces and nanostructures are inhomogeneous, containing two or more regions where the same magnetic species is found with a different number of magnetic and non-magnetic nearest neighbours. Since the magnetic properties of an interface are known to depend sensitively on nearest neighbour number and type [28], a full description of NI MSHG from an inhomogeneous interface must account for the contributions made by the different regions. It has been shown for NI MSHG that, through careful choice of sample alignment with respect to output polarization selection and the direction of applied magnetic field, the contribution from each region can be reduced to one *even* crystallographic tensor component and two *odd* magnetic tensor components [23].

The third rank crystallographic tensor components and fourth rank magnetic tensor components in Eq. (1) are expressed using the simplified notation $\chi_{ijk} \equiv ijk$ and $\chi_{ijkL} \equiv ijkL$, respectively, where the upper-case subscript $L$, describes the magnetization direction [14]. Systems of magnetic nanostructures grown by self-assembly at aligned atomic steps possess overall 1m symmetry. If the surface normal of the interface is in the $z$-direction, the normal to the single mirror-plane is in the $y$-direction along the steps, and the magnetization is in the $x$-direction ($M_x$), then the $y$-polarized NI MSHG intensity probes only the three tensor components, $yxy$, $yxxX$ and $yyyX$ [23]. For different magnetic regions, $n$, the dependence of the $y$-polarized NI MSHG intensity on $\varphi$, the angle between the input electric field vector and the $x$-direction, is given by

$$I_y(2\omega; \varphi, \pm M_X) \propto$$
$$|\sum_n yxy^{(n)} \sin 2\varphi \pm \{yxxX^{(n)} \cos^2 \varphi + yyyX^{(n)} \sin^2 \varphi\} M_X^{(n)}|^2 \quad (3)$$

MSHG thus offers the important diagnostic capability of exploiting the properties of the optical tensor components to identify different magnetization contributions from inhomogeneous interfaces and nanostructures, because the components will vary with the local atomic structure.

For centrosymmetric magnetic hysteresis loops, simplification of Eq. (3) occurs by eliminating the quadratic term using the Type II approach of Ref. [24]:

$$\Delta I_y^{\pm}(2\omega, \varphi, H) \equiv I^{\pm}(2\omega, \varphi, H) - I^{\mp}(2\omega, \varphi, -H)$$

$$\propto 4\sin 2\varphi \cos^2\varphi \sum_{n,n'} |\ yxy^{(n)}\ ||\ yxxX^{\ (n')}|\cos(\Delta\theta_{yx}^{nn'})M_X^{\pm(n')} \quad (4)$$

$$+ 4\sin 2\varphi \sin^2\varphi \sum_{n,n'} |\ yxy^{(n)}\ ||\ yyyX^{\ (n')}|\cos(\Delta\theta_{yy}^{nn'})M_X^{\pm(n')}$$

where $I^+$ has $H$ increasing from an initial negative value, etc., and $\Delta\theta_{yi}^{nn'} = \theta_{yxy}^{n} - \theta_{yiiX}^{n'}$, where $\theta$ are complex phase factors. This procedure removes all terms even in the magnetization, including magneto-elastic terms and cross-terms in the magnetization from different regions. It can be seen from Eq. (4) that choosing $\theta$ close to 0° or 90° will limit the magnetic contributions to a single component per region, $yxxX^{(n')}$ and $yyyX^{(n')}$, respectively. This much simplified equation can be fitted, for example, using sigmoidal magnetization loops. Only the magnetization depends on the applied magnetic field in Eq. (4), with the remaining terms determining the scale of the measured response. In the absence of other information, this prevents the strength of the magnetization being determined.

The full expression for the magnetic contrast or asymmetry is more complicated and, for $\theta$ close to 0° (and symmetry related angles), is given by:

$$A_{\varphi \approx 0} \equiv (I_y^+ - I_y^-)/(I_y^+ + I_y^-)$$

$$= 4\varphi \sum_{n,n'} |\ yxy^{(n)}||yxxX^{(n')}|\cos(\Delta\theta_{xx}^{nn'})M^{(n')} \div$$

$$(4\varphi^2 \sum_{n,n'} |yxy^{(n)}||yxy^{(n')}|\cos(\Delta\theta_{yxy}^{nn'}) \qquad (5)$$

$$+ \sum_{n,n'} |yxxX^{(n)}||yxxX^{(n')}|\cos(\Delta\theta_{yxxX}^{nn'})M^{(n)}M^{(n')})$$

Equation (5) shows that, as long as the temperature variation of the tensor components is similar, then substantial cancellation will occur in the asymmetry expression, allowing temperature-dependent behaviour to be probed.

## 4. Temperature-dependent MSHG from Fe nanostructures

Iron deposition on 1.4° vicinal W(110), and subsequent annealing, produces Fe nanostripes a single atom thick by self-assembly at the steps. The width of the stripes increases with further deposition until the (110) terraces are fully covered [2]. Figure 5 shows NI MSHG results, using unamplified 800 nm fs laser pulses, from 2 ML Fe, capped by 16 nm of Au, where two components can be identified. The Fe bilayer film nominally has two interfaces, W/Fe and Fe/Au, but calculations for Fe bilayers on W(110) show that the extended character of the electronic wavefunctions produces significant magnetic interactions extending three atomic layers into the substrate [29]: the Fe bilayer is part of a single interfacial region. RKKY indirect exchange coupling, mediated by the

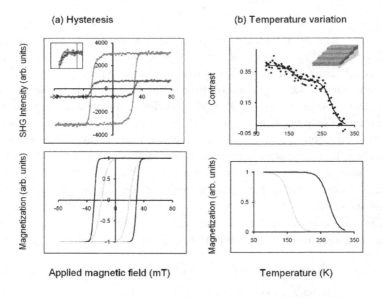

Figure 5. (a) NI MSHG y-polarized intensity difference, $\Delta I_y$ at 80 K and fits (solid lines), together with the resulting hysteresis curve (below), from Au-capped 2 ML Fe films grown on vicinal W(110): $\varphi \approx 0°$ (•), $\varphi \approx 90°$ (•). Inset: enlargement of normalized curves showing a small change in shape (b) Temperature variation of the MSHG contrast at $\varphi \approx 0°$, and simultaneous fits, together with extracted magnetization curves (below). Inset: schematic of 2 ML of Fe grown on the stepped W(110) substrate (after [27]). The Au capping layer is not shown.

conduction electrons of the substrate or capping layer, could also produce a second component. However, it has been shown previously for this capped system that effects of both dipolar and indirect exchange coupling between the stripes disappear, as expected, when the stripes overlap to form a contiguous film [30]. The second component is smaller and has a lower ferromagnetic transition temperature, consistent with the smaller number of exchange coupled neighbours of Fe atoms at the boundaries of the stripes [31].

Figure 6 shows NI MSHG results from 0.75 ML Fe, capped by 16 nm of Au, where two components can again be identified. The second component of

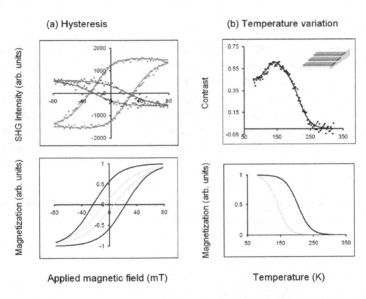

Figure 6. (a) NI MSHG y-polarized intensity difference, $\Delta I_y$ at 80 K and fits (solid lines), together with the resulting hysteresis curve (below), from Au-capped 0.75 ML Fe films grown on vicinal W(110): $\varphi \approx 0°$ (•), $\varphi \approx 90°$ (⊙). (b) Temperature variation of the MSHG contrast at $\varphi \approx 0°$, and simultaneous fits, together with extracted magnetization curves (below). Inset: schematic of 0.75 ML of Fe grown on the stepped W(110) substrate (after [27]). The Au capping layer is not shown.

the 0.75 ML sample also has a substantially lower transition temperature, consistent with boundary atoms. Previous work has identified only a small RKKY lateral interaction at this coverage [30]. The reversed loop of Figure 6 arises from a complex phase angle of greater than 90°, which has an additional consequence that the coherent contributions of the magnetic components partially destructively interfere, and thus the contrast increases when this contribution is switched off as the temperature increases. The coercivity and transition temperature data agree well with previous MOKE results [3, 30]. However, the MOKE studies provide no direct evidence of a second component in the magnetic response at sub-ML coverage. MOKE measures the total response from the step, edge and terrace atoms and, for this offcut and 0.75 ML Fe, boundary atoms will contribute only ~5% of the signal. The MOKE response is expected to be dominated by the terrace magnetization for this low angle offcut at these coverages. In contrast, SHG can be disproportionately sensitive to step structure [32], leading to the detection of boundary components of the magnetization.

## 5. Conclusion

MSHG offers the important diagnostic capability of exploiting the properties of the optical tensor components to identify different magnetization contributions from magnetic interfaces and nanostructures, because the tensor components vary with the local atomic structure. However, low symmetry systems, such as aligned magnetic nanostructures, are more difficult to characterize because of the large number of tensor components that may contribute to the signal. Normal incidence geometry simplifies the problem, while retaining sub-ML sensitivity.

## Acknowledgments

The author acknowledges the financial support of Science Foundation Ireland (SFI) provided under contract no. 05/RFP/PHY030.

## References

1. J. P. Velev, P. A. Dowben, E. Y. Tsymbal, S. J. Jenkins, and A. N. Caruso, *Surf. Sci. Rep.* **63**, 400 (2008).
2. J. Hauschild, H. J. Elmers, and U. Gradmann, *Phys. Rev.* **B57**, R677 (1998).
3. M. Pratzer and H. J. Elmers, *Phys. Rev.* **B67**, 94416 (2003).
4. P. Gambardella, A. Dallmeyer, K. Malti, M. C. Malagoli, W. Eberhardt, K. Kern, and C. Carbone, *Nature* **416**, 301 (2002).
5. P. Gambardella, A. Dallmeyer, K. Maiti, M. C. Malagoli, S. Rusponi, P. Ohresser, W. Eberhardt, C. Carbone, and K. Kern, *Phys. Rev. Lett.* **93**, 077203 (2004).
6. J. F. McGilp, *Prog. Surf. Sci.* **49**, 1 (1995).
7. Y. R. Shen, *Nature* **337**, 519 (1989).
8. S. B. Borisov and I. L. Lyubchanskii, *Optics and Spectroscopy* **61**, 801 (1986).
9. R.-P. Pan, H. D. Wei, and Y. R. Shen, *Phys. Rev.* **B39**, 1229 (1989).
10. J. Reif, C. Zink, C. M. Schneider, and J. Kirschner, *Phys. Rev. Lett.* **67**, 2878 (1991).

11. J.-W. Jeong, S.-C. Shin, I. L. Lyubchanskii, and V. N. Varyukhin, *Phys. Rev.* **B62**, 13455 (2000).

12. E. du Tremolet de Lacheisserie, *Phys. Rev.* **B51**, 15925 (1995).

13. Y. R. Shen, *The Principles of Nonlinear Optics*. New York: Wiley (1984).

14. A. Kirilyuk and T. Rasing, *J. Opt. Soc. Am. B* **22**, 148 (2005).

15. K. J. Veenstra, A. V. Petukhov, A. P. De Boer, and T. Rasing, *Phys. Rev.* **B58**, R16020 (1998).

16. M. Straub, R. Vollmer, and J. Kirschner, *Phys. Rev. Lett.* **77**, 743 (1996).

17. Q. Y. Jin, H. Regensburger, R. Vollmer, and J. Kirschner, *Phys. Rev. Lett.* **80**, 4056 (1998).

18. M. Nyvlt, F. Bisio, J. Franta, C. L. Gao, H. Petek, and J. Kirschner, *Phys. Rev. Lett.* **95**, 127201 (2005).

19. V. K. Valev, A. Kirilyuk, F. D. Longa, J. T. Kohlhepp, B. Koopmans, and T. Rasing, *Phys. Rev.* **B75**, 012401 (2007).

20. H. J. Choi, Z. Q. Qiu, J. Pearson, J. S. Jiang, D. Li, and S. D. Bader, *Phys. Rev.* **B57**, 12713 (1998).

21. H. J. Choi, R. K. Kawakami, E. J. Escorcia-Aparicio, Z. Q. Qiu, J. Pearson, J. S. Jiang, L. Dongqi, and S. D. Bader, *Phys. Rev. Lett.* **82**, 1947 (1999).

22. H. J. Choi, R. K. Kawakami, E. J. Escorcia-Aparicio, Z. Q. Qiu, J. Pearson, J. S. Jiang, D. Li, R. M. Osgood, III, and S. D. Bader, *J. Appl. Phys.* **85**, 4958 (1999).

23. L. Carroll and J. F. McGilp, *phys. stat. sol. (c)* **0**, 3046 (2003).

24. J. F. McGilp, L. Carroll, and K. Fleischer, *J. Phys.: Condens. Matter* **19**, 396002 (2007).

25. V. K. Valev, M. Gruyters, A. Kirilyuk, and T. Rasing, *phys. stat. sol. (b)* **242**, 3027 (2005).

26. L. Carroll, K. Fleischer, J. P. Cunniffe, and J. F. McGilp, *J. Phys.: Condens. Matter* **20**, 265002 (2008).

27. L. Carroll, J. P. Cunniffe, K. Fleischer, S. Ryan, and J. F. McGilp, *Phys. Rev. B*, to be published (2011).

28. H. J. Elmers, *Int. J. Mod. Phys.* **B9**, 3115 (1995).

29. X. Qian and W. Hubner, *Phys. Rev.* **B60**, 16192 (1999).

30. M. Pratzer and H. J. Elmers, *Phys. Rev.* **B66**, 033402 (2002).

31. R. Skomski and J. M. D. Coey, *Permanent Magnetism*. Bristol: Institute of Physics Publishing (1999).

32. J. R. Power, J. D. O'Mahony, S. Chandola, and J. F. McGilp, *Phys. Rev. Lett.* **75**, 1138 (1995).

# OPTICAL INVESTIGATIONS OF THE INTERFACE FORMATION BETWEEN ORGANIC MOLECULES AND SEMICONDUCTOR SURFACES

T. BRUHN[1], L. RIELE[1,2], B.-O. FIMLAND[3], N. ESSER[1,4] and P. VOGT[1]*

*1) Institut für Festkörperphysik, Technische Universität Berlin,
Hardenbergstr. 36, 10623 Berlin, Germany
* E-mail: patrick.vogt@tu-berlin.de*

*2) Dipartimento di Fisica, Universit di Roma Tor Vergata,
Via della Ricerca Scientifica 1, 00133 Rome, Italy*

*3) Department of Electronics and Telecomunications, Norwegian University of Science
and Technology
NO-7491 Trondheim, Norway*

*4) Leibniz-Institut für Analytische Wissenschaften - ISAS e.V.
Albert-Einstein-Str. 12, 12489 Berlin, Germany*

We report on the characterization of organic layers adsorbed on semiconductor surfaces by means of reflectance anisotropy spectroscopy (RAS). At different organic adsorbate layers on GaAs(001) surfaces we show that RAS measurements are suitable to distinguish between physisorption and chemisorption of organic molecules on semiconductur surfaces. We also show that molecular layers can contribute to RAS signatures with new anisotropies due to intramolecular optical absorption. These molecular anisotropies can be utilzed to determine molecular orientation or to monitor the growth of organic layers directly for sub-monolayers as well as for thicker layers.

*Keywords*: RAS, organic layers, in-situ monitoring, organic molecules, adsorption

## 1. Introduction

Interfaces between organic materials and semiconductor surfaces represent a rather new field of research that meets applications in organic/inorganic hybrid devices from OLEDs and OFETs to medical techniques and biosensors.[1–3] Within this field one of the key issues is the controlled preparation of ultra-thin organic layers from single monolayers to thicker layers in the range of several nanometers. For such a control it is inevitable to have an experimental tool that provides a non-destructive in-situ monitoring of the organic layers growth under different atmospheric conditions. Another challenge is present in the characterization of the properties of adsorbed organic layers in terms of the molecular ordering or orientation within the layer as also the characterization of the bonding configuration in the interface to the semiconductor material.

So far, the properties of organic layers on semiconductor surfaces have mainly been studied by methods that apply x-ray (XPS, SXPS, XRD...) or electron (ELS, AES, LEED, RHEED,...) radiation onto the organic material. Electron or x-ray radiation could, however, be shown to interact significantly with molecular layers leading either to a modification of their electronic and structural properties or even to a destruction or desorption of the adsorbed molecules.[4]

Optical techniques, on the contrary, have proven to allow a profound, non-destructive characterization of surfaces and interfaces of the investigated material. Over many years particularly reflectance anisotropy spectroscopy (RAS) has provided substantial insights into the properties of semiconductor surfaces and helped significantly in understanding the surface structures of III-V semiconductors (GaAs, GaP, InP, InGaP).[5]

RAS helped to understand surface formations where other experimental methods do not allow for direct access e.g. surface stabilization by hydrogen like for example for the clarification of the hydrogen-stabilized InP$(2 \times 1)$ surface, demonstrating its potential for the characterization of adsorbates on semiconductor surfaces.[6] For InP $(2 \times 4)$ it could be demonstrated that surface related RAS transitions can be explained by transitions between electronic states localized at surface structures e.g. dangling bonds or back bonds of surface atoms.[7–9]

It seems reasonable to transfer this potential of RAS to the analysis of bare surface structures to interfaces between semiconductor surfaces and organic molecules. In this case it can be expected to find changes of surface related features of the clean surfaces upon molecule adsorption as well as direct contributions from the molecules e.g. intra-molecular transitions. However, the optical transitions of small organic molecules are often found in the spectral range above 5 eV, e.g. the HOMO-LUMO gap of pyrrole is 5.9 eV and the HOMO-LUMO (highest occupied molecular orbital - lowest unoccupied molecular orbital) gap of benzene is 6.5 eV. For the characterization of optical anisotropies within organic layers it is therefore required to modify the RAS setup for an operation in the UV spectral range. Here, we investigate the adsorption of several small cyclic organic molecules (cyclopentene, benzene, pyrrole,... lead-phthalocyanine) on reconstructed GaAs(001) surfaces by in-situ RAS. The monitoring of the adsorption process allows the preparation of organic sub-monolayer and provides insights into the bonding configuration (physisorption/chemisorption, orientation,...) of the adsorbed molecules.

In order to demonstrate the applicability of RAS for the interface analysis of semiconductor/molecule structures we will discuss three different examples. In A) we compare the influence of molecular physisorption and chemisorption on the RAS signatures of semiconductor surfaces. In B) we investigate the molecule-specific RAS signatures of organic monolayers. In C) we demonstrate the potential of RAS for the characterization of thicker organic layers for the in-situ analysis of the growth behavior of organic layers.

As a substrate for our adsorption experiments we used reconstructed GaAs(001) surfaces. The As-rich $c(4 \times 4)$ surface consists of As-As dimer triplets which are bonded to a second layer of As. The Ga-rich $(4 \times 2)$ surface exhibits As as well as Ga atoms in the outermost layer with a Ga-Ga sub-surface dimer. The ball-and-stick models of the respective surface reconstructions are shown in Fig. 1 according to Schmidt et al.[8] The molecules deposited onto these GaAs surfaces were cyclopentene, pyrrole, 3-pyrroline and lead-phthalocyanine (PbPc) as depicted in Fig. 2.

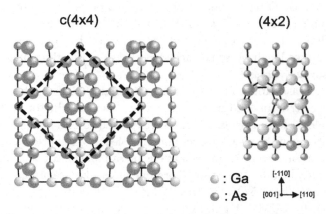

Fig. 1. Ball-and-stick models of the GaAs(001)-$c(4 \times 4)$ and $(4 \times 2)$ reconstruction according to Schmidt et al.[8]

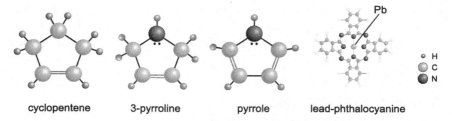

cyclopentene          3-pyrroline          pyrrole          lead-phthalocyanine

Fig. 2.  Ball-and-stick structure models of the organic molecules discussed within this paper. While cyclopentene and 3-pyrroline contain only one C=C bond and are hence non-aromatic, pyrrole and lead-phthalocyanine (PbPc) are aromatic molecules with delocalized $\pi$-electrons.

## 2. Experimental

For the experiments MBE grown and As-capped GaAs samples were used with nominal Si doping of below $n = 5 \times 10^{17} cm^{-3}$. Reconstructed surfaces were obtained by decapping at $330°C$ under UHV condictions and further annealing.[10] The base pressure of the UHV system was below $2 \times 10^{-10}$ mbar throughout all experiments. Cyclopentene, pyrrole and 3-pyrroline were purchased from Sigma-Aldrich with a purity of 96% (cyclopentene) and 99% (pyrrole, 3-pyrroline), respectively. For the deposition of cyclopentene, pyrrole and 3-pyrroline, the molecules were dosed into the UHV system from gas-phase via a dosing valve. The purity of the molecules led into the UHV system was checked by QMS before the deposition process. During the deposition all filaments were switched off in order to avoid decomposition of the molecules. The partial pressure during the deposition procedure lay between $1 \times 10^{-7}$ and $1 \times 10^{-4}$ mbar. All values concerning the amount of molecules deposited in the different cases are given in Langmuir. PbPc layers were deposited via an organic effusion cell (MBE components) at a temperature of $350°C$. Details on the PbPc deposition are given elsewhere.[11] For the RAS measurements, two different setups were used. The first one was a standard RAS setup according to Aspnes et al. operating in a photon energy range from 1.5 to 5.5 eV.[12,13] In the second setup, a deuterium lamp was used as light source. The design of the setup and the optical path was in principle similar to the visible RAS setup but all optical components (windows, coatings etc.) were adapted to the UV application and the entire setup as well as the single optical components was purged with nitrogen. The setup was attached to a UHV vessel via a low-strain $CaF_2$ window. More details can be found in Ref. 14. This second setup provided an operation range between 3.0 eV and 8.5 eV. The measurements showed excellent agreement in the spectral range between 3.0 and 5.5 eV accessible to both setups. The spectra presented here from 1.5 to 8 eV in one graph have therefore been obtained by combining the results of the two different setups.

## 3. Results and Discussion

### 3.1. *A) RAS monitoring of molecular physisorption or chemisorption*

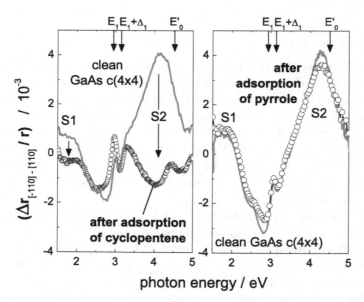

Fig. 3. RAS signatures of GaAs(001)-$c(4 \times 4)$ surfaces after adsorption of cyclopentene (left) and pyrrole (right). The chemisorption of cyclopentene (left) causes a significant modification of the RAS signature at the surface transitions S1 (below 2 eV) and S2 (4 eV). On the contrary, the physisorption of pyrrole (right) does not cause any significant modification at the surface transitions due to the weak molecule-surface interaction.

Fig.3 shows the RAS spectra of the GaAs(001)-$c(4 \times 4)$ surface before and after the adsorption of cyclopentene (left) and pyrrole (right). The non-aromatic molecule cyclopentene was found to chemisorb on the $c(4 \times 4)$ surface via covalent bonds from the C atoms involved in the intramolecular C=C bond to the surface As atoms.[15-17] In the RAS signature this chemisorption causes a significant modification of the surface features S1 below 2 eV and S2 around 4 eV that result from transitions localized at the surface As-As dimers.[8,9] The observed changes in the RAS signature are thus in agreement with structural and hence electronic changes at the As-As dimer upon chemisorption of the cyclopentene molecule.

In the RAS spectra after pyrrole adsorption, however, no significant modification of the surface RAS signature can be observed. As discussed elsewhere[18] this is a clear indication that the aromatic molecule pyrrole adsorbs on the $c(4 \times 4)$ surface only via a weak physisorption without the formation of covalent bonds between molecule and surface.

This means that the surface RAS signature is sensitive to the adsorption of organic molecules and allows to distinguish between chemisorption and physisorption processes during the interface formation. However, this modification of the surface RAS signature is not specific to different molecular species because it only monitors the molecule adsorption indirectly via the modification of the surface due to the surface-molecule interaction. As long as the bonding is formed to the As-As dimer the changes of the RAS signature are very similar. This can be seen for example for the adsorption of 1,4-cyclohexadiene on the $c(4 \times 4)$ surface which causes essentially the same modification of the surface transitions S1 (below 2 eV) and S2 (4 eV) as cyclopentene as discussed elsewhere.[16,17,19]

## 3.2. B) RAS monitoring of intra-molecular transitions

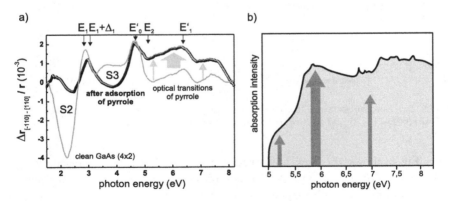

Fig. 4. a) RAS spectra recorded before (grey line) and after (open circles) adsorption of pyrrole. The anisotropies S2 and S3 related to surface transitions are reduced upon adsorption and new anisotropies can be observed between 5 and 8 eV.

For a direct measurement of anisotropies caused by the adsorbed molecules, the spectral range of the RAS was extended to the spectral range from 1.5 to 8 eV. In this range intra-molecular transitions are found, e. g. of pyrrole at 5.2, 5.9 and 7.2 eV.[20,21] A measurement of the optical absorption of pyrrole is depicted in Fig. 4 b) according to Mullen et al.[22] showing an inset of the absorption at 5 eV, a maximum at 5.9 eV and several broad absorption bands above 6.5 eV. In Fig. 4 a) the adsorption of pyrrole on the Ga-rich GaAs(001)-(4 × 2) surface is investigated by RAS. It can be seen, that the surface anisotropies S2 (2.2 eV) and S3 (3.5 eV) are reduced significantly upon pyrrole adsorption indicating a chemisorption process.[23] Additionally it can be seen that new anisotropies arise upon adsorption of pyrrole between 5 and 8 eV. Within this spectral range also the optical transitions of pyrrole were identified with a broad absorption maximum around the HOMO-LUMO gap at 5.86 eV[20,21] as depicted in b). The new anisotropies between 5 and 8 eV can therefore be explained by an anisotropic optical absorption due to intra-molecular transitions. This in turn points towards an anisotropic arrangement of the adsorbed molecules, for example due to a preferential orientation.

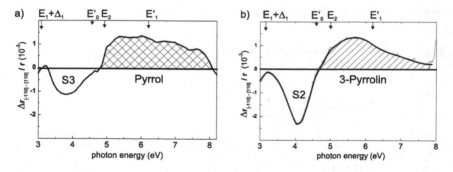

Fig. 5. New anisotropies (ΔRAS) resulting from the adsorption of a) pyrrole on GaAs(001)-(4 × 2) and b) 3-pyrroline on GaAs(001)-$c$(4 × 4) surfaces.

Fig. 5 shows the new anisotropies between 3 and 8 eV caused by the adsorption of pyrrole and 3-pyrroline on GaAs(001) surfaces. The spectra were obtained by subtracting the spectrum of the clean surface from the spectrum measured after molecule adsorption. The comparison shows that the different molecules also cause different new anisotropies in the RAS spectra. While pyrrole adsorption causes broad new anisotropies between 5 and 8 eV without a clear maximum the adsorption of 3-pyrroline leads to new anisotropies with a pronounced maximum at 5.9 eV. As discussed above the anisotropies observed for the pyrrole layer can be attributed to intra-molecular transitions. For gas-phase 3-pyrroline no optical absorption spectra are available so far. The comparison of Fig. 5 a) and b) however shows that the new anisotropies are specific for different molecular species correlating to the absorption bands of the adsorbed molecule.

### 3.3. C) Thicker organic layers and growth behavior

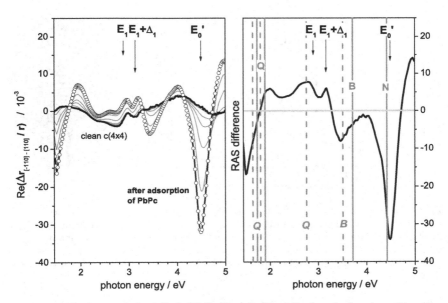

Fig. 6.   RAS spectra of clean GaAs(001)-$c(4 \times 4)$ (black line) and after deposition of a 22 nm of PbPc (open circles). b) Difference spectra between the RAS after adsorption of the 20 nm PbPc and the clean $c(4 \times 4)$. Positions of molecular absorption bands of PbPc in vapor phase (solid grey lines) and on quartz glass (dashed grey lines). The new anisotropies after PbPc can be attributed to intra-molecular transitions.

Apart from the analysis of only one-monolayer thick organic layers on semiconductor surfaces RAS allows also to follow and analyze the growth of thicker layers. As an example Fig. 6 shows a series of RAS spectra between 1.5 and 6.5 eV recorded during the growth of a PbPc film on GaAs(001)-$c(4 \times 4)$. The last spectrum refers to a layer thickness of approximately 20 nm. The spectra show new anisotropies at the Q (1.6 eV), B (3.5 eV) and N (4.4 eV) optical absorption band of PbPc band which scale with the growth of the PbPc film. The appearance of new anisotropies from the adsorbed molecular layer indicates, that the molecules within the PbPc layer arrange in such a way that the optical transitions are not isotropically ordered. Since the PbPc molecule is in-plane isotropic such an anisotropic arrangement can only result from a preferential ordering of the molecules within the molecular film within a tilting angle of the molecular plane relative to the surface. This means that RAS allows to identify a tilting of the moleculas with respect to the substrate surface.

a)                                              b)

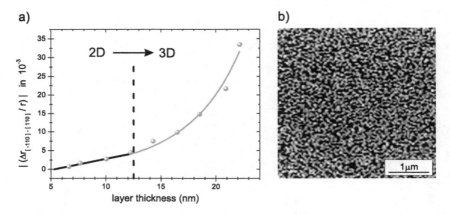

Fig. 7. a) Transient of the RAS intensity at 4.5 eV plottet versus the thickness of the PbPc layer deposited on a GaAs $c(4 \times 4)$ substrate. The linear behavior up to 10 nm indicates a 2D layer-by-layer growth which is followed by a 2D to 3D transition indicated by the exponential behavior above 10 nm layer thickness. b) AFM image $(4 \times 4 \mu m)$ of a 20 nm thick PbPc film on GaAs $c(4 \times 4)$ showing a granular 3D structure of the PbPc layer.

In Fig. 7 a) the RAS intensity at 4.5 eV is plotted versus the deposition time. It can be seen that the intensity of the anisotropy increases with increasing deposition. This indicates that the growing layers keep the anisotropic arrangement of the first PbPc monolayer. For the latter it was shown that the atomic structure of the initial GaAs $c(4 \times 4)$ surface is responsible for the anisotropic arrangement of the adsorbed PbPc molecules. In contrast to this, it was shown for PbPc layers deposited on the $(2 \times 4)$ reconstructed GaAs(001) surface that no new anisotropies due to intramolecular transitions can be observed in RAS measurements because of an isotropic molecular arrangement within the adsorbed PbPc layer.[11]

The linear increase of the anisotropy in Fig. 7 a) for the first 10 monolayers indicates a 2D layer-by-layer growth in this range while the exponential increase above 10 monolayers indicates a 2D to 3D growth transition of the PbPc layer. Such a 2D to 3D transition was also confirmed by AFM measurements that revealed a granular structure of the 20 nm thick PbPc layer as depicted in Fig. 7 b).

The observation of anisotropies due to intra-molecular transitions, therefore, allows the determination of the growth mode of molecular layers and the determination of a molecular ordering within such organic layers.

## 4. Summary

We have shown that RAS measurements allow a profound characterization of the adsorption processes of organic molecules on III-V semiconductor surfaces. Via the adsorption-induced modification of anisotropies resulting from surface transitions we could identify whether molecules physisorb or chemisorb at the surface. We have also extended the spectral range of the RAS to the UV range in order to provide a direct observation of anisotropies from intra-molecular transitions allowing the identification of the adsorbed molecular species. We have also applied RAS measurements for the characterization of thicker organic layers and demonstrated its potential for the identification of the growth modes and molecular orientation within organic layers. For the future there are several promising challenges for the application of RAS on organic adsorbate layers. For example polar measurements with different angles between the surface crystal axes and the polarization state of light could give deeper insights into the exact molecular orientation. Furthermore 2D RAS at fixed photon energy could allow the direct observation of the formation of domains within organic layers.

## 5. Acknowledgement

We would like to thank E. Speiser and C. Cobet for their help with the installation of the UV RAS.

## References

1. R. J. Hamers, *Annual Review of Analytical Chemistry* **1**, 707 (2008).
2. C. Sanchez, H. Arribart and M. M. Giraud Guille, *Nature Materials* **4**, 277 (2005).
3. M. Stutzmann, J. A. Garrido, M. Eickhoff and M. S. Brandt, *phys. stat. sol. (a)* **203**, 3424 (2006).
4. C. D. MacPherson and C. D. Leung, *Surf. Sci.* **324**, 202 (1995).
5. P. Weightman, D. S. Martin, R. J. Cole and T. Farrell, *Rep. Prog. Phys.* **68**, 1251 (2005).

58

6. P. Vogt, T. Hannappel, S. Visbeck, K. Knorr, N. Esser and W. Richter, *Phys. Rev. B* **60**, R5117 (1999).

7. W. Schmidt, N. Esser, A. M. Frisch, P. Vogt, J. Bernholc, F. Bechstedt, M. Zorn, T. Hannappel, S. Visbeck, F. Willig and W. Richter, *Phys. Rev. B* **61**, R16335 (2000).

8. W. G. Schmidt, F. Bechstedt, K. Fleischer, C. Cobet, N. Esser, W. Richter, J. Bernholc and G. Onida, *phys. stat. sol. (a)* **188**, 1401 (2001).

9. W. G. Schmidt, F. Bechstedt and J. Bernholc, *Appl. Sur. Sci.* **190**, 264 (2002).

10. U. Resch-Esser, N. Esser, D. T. Wang, M. Kuball, J. Zegenhagen, B. O. Fimland and W. Richter, *Surf. Sci.* **352**, 71 (1996).

11. L. Riele, T. Bruhn, V. Rackwitz, R. Passmann, B. O. Fimland, N. Esser and P. Vogt, *Phys. Rev. B* **submitted** (2011).

12. D. E. Aspnes, J. P. Harbison, A. A. Studna and L. T. Florez, *Appl. Phys. Lett.* **52**, 27 (1988).

13. W. Richter, *Phil. Trans. R. Soc. A* **344**, 453 (1993).

14. E. Speiser, C. Cobet, T. Bruhn, M. Ewald, P. Vogt, W. Richter and N. Esser, *Appl. Phys. Lett.* **to be published** (2011).

15. R. Passmann, M. Kropp, T. Bruhn, B. O. Fimland, F. L. Bloom, A. C. Gossard, W. Richter, N. Esser and P. Vogt, *Appl. Phys. A* **87**, 469 (2007).

16. R. Passmann, T. Bruhn, T. A. Nilsen, B. O. Fimland, M. Kneissl, N. Esser and P. Vogt, *phys. stat. sol. (b)* **246**, 1504 (2009).

17. T. Bruhn, R. Passmann, B. O. Fimland, M. Kneissl, N. Esser and P. Vogt, *phys. stat. sol. (b)* **247**, 1941 (2010).

18. T. Bruhn, B. O. Fimland, M. Kneissl, N. Esser and P. Vogt, *Phys. Rev. B* **to be published** (2011).

19. T. Bruhn, B. O. Fimland, M. Kneissl, N. Esser and P. Vogt, *Phys. Rev. B* **83**, 045307 (2011).

20. O. Christiansen, J. Gauss, J. F. Stanton and P. Jorgensen, *J. Chem. Phys.* **111**, 525 (1999).

21. M. H. Palmer, I. C. Walker and M. F. Guest, *Chem. Phys.* **238**, 179 (1998).

22. P. A. Mullen and M. K. Orloff, *J. Chem. Phys.* **51**, 2276 (2003).

23. T. Bruhn, M. Ewald, B. O. Fimland, M. Kneissl, N. Esser and P. Vogt, *J. Nanopart. Res.* **in print** (2011).

# REFLECTION ANISOTROPY SPECTROSCOPY STUDIES OF THIOLATE/METAL INTERFACES

D. S. MARTIN

*Department of Physics and Surface Science Research Centre,*
*University of Liverpool, L69 7ZE, England, UK*
*E-mail: David.Martin@liverpool.ac.uk*

Recent work on the application of reflection anisotropy spectroscopy (RAS) to thiolate/metal interfaces is reviewed. The thiolate-metal surface bond is a key interaction in the attachment of alkanethiols and thiol-derivatised molecules to metal surfaces under ultra-high vacuum conditions and at the solid/liquid interface. Characteristic RAS signatures are obtained from thiolate/Cu(110) and thiolate/Au(110) surfaces, and these signatures can be simulated using three- and four-phase models.

## 1. Introduction

Reflection anisotropy spectroscopy (RAS) is a linear optical probe of surfaces and interfaces. RAS belongs to the "epioptics" family of linear and non-linear spectroscopies[1] which exploit the difference in symmetry between the surface or interface region and the bulk material to achieve surface sensitivity. When applied to cubic crystals, RAS becomes surface sensitive since the optical response of the bulk crystal is isotropic, and the signal arises from the lower symmetry of the surface. In a RAS experiment, plane-polarized light illuminates the surface of interest at near-normal incidence and the difference in reflection between two orthogonal directions in the surface plane is measured.[2] For the noble metal fcc(110) surfaces the RAS measurement is defined as the difference in reflectance ($\Delta r$) normalised to the mean reflectance ($r$)

$$\frac{\Delta r}{r} = \frac{2(r_{[1\bar{1}0]} - r_{[001]})}{r_{[1\bar{1}0]} + r_{[001]}} \tag{1}$$

where the complex Fresnel reflectance amplitudes $r$ are different for $[1\bar{1}0]$ and $[001]$ surface directions. The experimental aspects of RAS have been reviewed.[1-9] An important advantage of RAS over electron-based surface spectroscopies is the ability of RAS to probe surfaces in a range of gas pressures from ultra-high vacuum (UHV) to ambient and higher pressure environments, and the ability to investigate solid/liquid interfaces.

The adsorption and assembly of molecules into two-dimensional monolayers (ML) on surfaces is important in many fields and applications. The thiolate-metal surface bond is a key interaction that finds widespread use in the attachment of alkanethiols and thiol-derivatised molecules to metal surfaces under aqueous or UHV environments. The structural anisotropy of fcc(110) surfaces can be exploited as a template to influence the two-dimensional ordering in the molecular layer. Here, a brief review is given of some recent results of using RAS to probe the optical properties of thiolate/Cu(110) and thiolate/Au(110) surfaces. The thiolate molecules chosen for study have no absorption bands in the range of the RAS spectrometer (1.5 to 5.0 eV) and so the observed changes in RAS upon adsorption are primarily due to changes of the substrate surface optical response, resulting from electronic and/or morphological re-structuring.

## 2. Thiolate/Cu(110) surfaces

In recent work,[10] we have investigated the optical properties of thiolate/Cu(110) surfaces created by the adsorption of (i) the simplest alkanethiol, methanethiol [$CH_3SH$] and (ii) the amino acid L-cysteine [$HS-CH_2-CH(NH_2)-COOH$]. For both these molecules the S–H bond is broken upon adsorption onto Cu(110) and the molecules adsorb as a thiolate species. In the case of L-cysteine there is also an amine and carboxylic acid group that have the potential to interact with the Cu surface. By comparing the results of L-cysteine/Cu with $CH_3S$/Cu, some insight can be offered into the importance of the thiolate interaction in the L-cysteine/Cu system.

The RAS profile obtained following exposure of Cu(110) to ~3 Langmuir (L) (~0.3 ML) of methanethiol is shown by the open circles in Fig. 1a. This spectrum is very similar to the spectrum observed for high-coverage L-cysteine/Cu(110) shown by the filled triangles in Fig. 1a. The similarity suggests a similar bonding interaction and surface structure for L-cysteine/Cu(110) and $CH_3S$/Cu(110). For both adsorbate systems, a $c(2 \times 2)$ low-energy electron diffraction (LEED) pattern is observed; methanethiolate has broad diffuse spots at the half-order positions, L-cysteine has half-order spots split into doublets along the real space [$1\bar{1}0$] direction suggesting the presence of domains. The RAS response at 4.2 eV for L-cysteine shows a well defined peak that is different from the response observed for methanethiolate at the same energy (Fig. 1a). The peak indi-

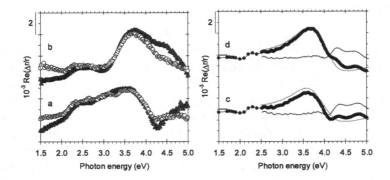

Fig. 1. RA spectra of (a) CH₃S/Cu(110) (open circles) and L-cysteine/Cu(110) (triangles), (b) S/Cu(110)-c(8 × 2)-type structure formed by heating CH₃S/Cu(110) (open circles) or formed by heating L-cysteine/Cu(110) (triangles). The simulated RA spectra of (c) and (d) simulate the spectra of (a) and (b) respectively. Dark and light lines correspond to contributions from the first and second terms of Eq. (2), respectively.

cates the presence of a carboxylate interaction for L-cysteine/Cu(110), as previous work has found that the carboxylate-Cu interaction contributes a characteristic RAS intensity at 4.2 eV.[11] Thus the RAS data of Fig. 1a indicate that L-cysteine is interacting with the Cu surface via a thiolate and a carboxylate linkage. Mateo Marti et al.[12] have investigated the adsorption of S-cysteine on Cu(110) using reflection absorption infrared spectroscopy and x-ray photoelectron spectroscopy (XPS). Their results indicate that the molecule interacts with the Cu surface via a thiolate and a carboxylate linkage, consistent with the RAS results.

It was found that a thermal treatment of heating to 800 K and subsequent cooling to room temperature produced similar results for both CH₃S/Cu and L-cysteine/Cu.[10] The RAS profile of Fig. 1b (open circles) is obtained following the thermal treatment applied to a high coverage (exposure range 40-70 L) CH₃S/Cu surface. The LEED pattern of this surface was of an incomplete c(8 × 2)-type pattern similar to that reported by Domange and Oudar for Cu(110) following exposure to H₂S.[13] The same LEED pattern and similar RAS profile are observed (Fig. 1b triangles) following the same thermal treatment applied to the saturated L-cysteine/Cu(110) surface. XPS results[10,14] have shown that the heating procedure for both CH₃S/Cu(110) and L-cysteine/Cu(110) surfaces results in the break-up of the molecules to leave only chemisorbed S on the surface. Thus the RAS profile of Fig. 1b is an optical signature of the S/Cu(110)-c(8 × 2)-type structure.

## 3. Simulating the data - a four-phase model

To simulate the RAS response of molecule/metal interfaces, the three-phase model of McIntyre and Aspnes[15] that is commonly used to simulate clean surfaces can be expanded to incorporate a fourth "overlayer" phase to represent the adsorbate. As illustrated in Fig. 2 each phase has its own complex dielectric function $\epsilon$, i.e. vacuum ($\epsilon_1 = 1$), a biaxial anisotropic overlayer ($\Delta\epsilon_2 = \epsilon_2^x - \epsilon_2^y$) and surface ($\Delta\epsilon_3$) and an isotropic bulk ($\epsilon_4$). The vacuum and bulk phases are semi-infinite whereas the overlayer and surface phases have thickness $d \ll \lambda$ the wavelength of light to satisfy the thin-film limit.[15] The reflection coefficients and $\Delta r/r$ for normal incidence reflection from the stacked four-phase system are then determined. The RAS response of the four-phase system is given by[10]

$$\frac{\Delta r}{r} = \frac{4\pi i}{\lambda} \left[ \frac{d_3(\Delta E_g - i\Delta\Gamma)}{\epsilon_4 - 1} \frac{d\epsilon_b}{dE} + \frac{d_2 \Delta\epsilon_2}{\epsilon_4 - 1} \right]. \tag{2}$$

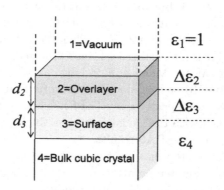

Fig. 2.   Four phase model used to simulate the RAS response of overlayer systems.

To simulate the RAS of clean Cu(110) between 3.0 to 5.5 eV, the region dominated by transitions near the L symmetry point, $\Delta\epsilon_3$ is based on the energy derivative of the bulk dielectric function, $d\epsilon_b/dE$.[16,17] The anisotropic Cu(110) surface is assumed to perturb the electronic structure of the subsurface region, causing different band narrowing along [1$\bar{1}$0] and [001] leading to differences in gap energies $\Delta E_g$ and linewidths $\Delta\Gamma$ of interband transitions at L between $x$ and $y$ polarizations. A three-phase model (no overlayer) with values of $\Delta E_g = 0.1$ and $\Delta\Gamma = 0$ produce a simulated spectrum that is in good agreement with experiment, with peaks at 4.2 and 4.9 eV occuring close to the experimentally observed peaks.[16]

The real part of Eq. 2 (using $\epsilon = \epsilon' - i\epsilon''$) can be used to simulate the experimental RAS results $(\mathrm{Re}(\Delta r/r))$ of the thiolate/Cu(110) systems. The overlayer phase $\Delta\epsilon_2$ is based upon a single oscillator of energy $\omega_t$, strength $S$ and width $\gamma$ occurring for one of the RAS polarizations. It is found that an oscillator for $y$ polarization gives simulated RAS profiles with sign in agreement with experiment. For this case, the oscillator is described by[18]

$$\epsilon_2^y = 1 + \frac{S/\pi}{\omega_t - \omega + i\gamma/2}; \epsilon_2^x = 1. \tag{3}$$

Thickness $d_2$ and $d_3$ are both set to 1 nm, leaving $\omega_t$, $S$ and $\gamma$ of the oscillator and $\Delta E_g$, $\Delta\Gamma$ of the surface as variables. The dielectric properties of Cu $(\epsilon_b)$ are obtained from tabulated data below 2.5 eV,[19] and above 2.5 eV higher resolution data on single crystal Cu is available.[20]

Simulations of the thiolate/Cu(110) results are shown in Fig. 1c–d. The parameters used in the simulations are shown in Table 1. The simulations reproduce the main features of the experimental spectra (Fig. 1) and the single overlayer transition of similar energy for the simulations implies a common origin, i.e. the bonding interaction at the interface. A limitation of the four-phase model is that as the layer contributions are additive (Eq. 2), it is difficult to distinguish between separate contributions from the surface and overlayer, and more than one contribution from the same layer. An oscillator of the form of Eq. 3 is typically used in simulating transitions localised at the interface e.g. transitions between surface states[18] whereas adsorbate-induced effects in the surface layers generally produce derivative-like RA spectra. The resulting simulations with these assumptions show good agreement with experiment (Fig. 1), however, these assumptions may not always be valid and as such the 3.8 eV transition cannot be unambiguously assigned directly to the thiolate-Cu bond. The profiles in Fig. 1 are characteristic optical signatures of the system which originate in the interfacial region of the topmost Cu layers and the adsorbate overlayer.

Results of first principles calculations[21] on the structural and electronic properties of $CH_3S$/Cu(110)-$c(2 \times 2)$ show that the bonding involves a strong hybridization between $p$ orbitals of the molecular S and $d$ states of the Cu surface. This leads to an increase, relative to clean Cu, in the density of states in the Cu surface layer of the $CH_3S$/Cu interface between 3 to 4 eV binding energy[21] and a bonding state is found at $\sim$3.5 eV binding energy. The energy, $\sim$3.8 eV, of the broad transition in the overlayer used in the simulations of Fig. 1c–d is consistent with the theoretical results.

Table 1. Values used in the simulations of the thiolate/Cu(110) and thiolate/Au(110) surfaces. The range refers to the energy region of the RA spectrum being simulated.

| Structure | Range/eV | $\Delta E_g$/eV | $\Delta\Gamma$/eV | $\omega_t$/eV | $S$ | $\gamma$/eV |
|---|---|---|---|---|---|---|
| Thiolate/Cu-$c(2 \times 2)$ [Fig. 1c] | 1.5-2.5 | 0 | 0.01 | - | - | - |
| | 2.5-5.5 | 0.01 | 0.03 | 3.8 | 0.7 | 1.0 |
| Thiolate/Cu-$c(8 \times 2)$-type [Fig. 1d] | 1.5-2.5 | 0 | 0.01 | - | - | - |
| | 2.5-5.5 | -0.01 | 0.03 | 3.9 | 0.9 | 1.0 |
| Au(110)-$(1 \times 2)$ [Fig. 4] | 1.5-3.0 | 0.02 | 0 | 1.8 | 2.0 | 0.6 |
| | 3.0-5.0 | 0.2 | -0.05 | - | - | - |
| Thiolate/Au(110) [Fig. 4] | 1.5-3.0 | 0.04 | 0 | - | - | - |
| | 3.0-5.0 | 0.1 | -0.05 | - | - | - |

## 4. Thiolate/Au(110) surfaces

The Au(110) surface is relatively straightforward to prepare in UHV and for electrochemical studies at the solid/liquid interface, allowing a comparative investigation of the properties of electrochemical and vacuum-deposited monolayers. The adsorption of L-cysteine onto Au(110) in UHV and under liquid has been studied[22] and the results are now summarised. The thiol group of L-cysteine again plays an important role: L-cysteine is known to adsorb onto Au surfaces via a covalent thiolate linkage.[23]

The Au(110) surface at room temperature in UHV adopts the $(1 \times 2)$ missing row reconstruction where every second $[1\bar{1}0]$ row of atoms in the top layer is absent. The RAS profile of this surface, shown by the open circles in the left panel of Fig. 3, is characterised by positive peaks at 1.8 eV and 4.5 eV and negative peaks at 2.5 eV and 3.5 eV. These RAS features are thought to arise from bulk band transitions near the L point that are modified by the presence of the anisotropic surface,[24] similar to Cu(110). The 1.8 eV peak (Fig. 3) may have a contribution from transitions between surface states at the $\bar{\Gamma}$ symmetry point. The RA spectrum of the vacuum deposited L-cysteine surface at saturation coverage is shown by the filled circles in the left panel of Fig. 3. As L-cysteine is deposited onto the clean Au(110) surface, three changes to the RA spectrum can be identified: (i) a reduction in the RAS peak at 1.8 eV, (ii) an increase in intensity of peak at 2.5 eV, and (iii) a decrease in intensity of the peak at 3.5 eV. The change around 1.8 eV upon adsorption is consistent with the destruction of transitions between surface states.

The RAS profile of Au(110)-$(1 \times 2)$ at a potential of 0 V in phosphate buffer solution (pH 7) is shown by the open circles in the right panel of Fig. 3. The spectrum is similar to that of the clean surface prepared in UHV. Following introduction of L-cysteine into the electrochemical cell (0.1mM

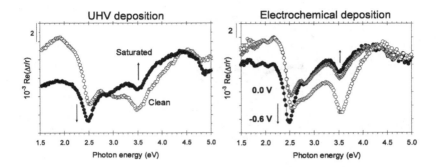

Fig. 3. Comparison of RA spectra of L-cysteine/Au(110) deposited in UHV (left panel) and in an electrochemical cell (right panel).

L-cysteine/0.1 M phosphate buffer), RA spectra obtained at potentials of 0 V and -0.6 V are shown in Fig. 3. The RA spectrum for -0.6 V (filled circles, right panel) is similar to that observed for the L-cysteine saturated surface in UHV (filled circles, left panel) and suggests similar adsorption characteristics. Comparing the RA spectra of L-cysteine/Au(110) at 0 V and at -0.6 V, a similar response around 3.5 eV is observed, however, the behaviour at 2.5 eV is different. The intensity of the peak at 2.5 eV is found to increase with increasingly negative potentials. The negative potential should attract the amine group to the Au surface and repel the carboxylate group and in doing so produce a similar orientation to that found in UHV.[23]

Results of another thiolate molecule, decanethiol, on Au(110)[25] have shown that at saturation coverage, a RAS profile similar to the -0.6 V L-cysteine/Au(110) profile is observed. Decanethiol molecules attach to Au surfaces as thiolates and the similarity in RAS profiles of L-cysteine/Au (Fig. 3) and decanethiolate/Au[25] indicates that the thiol group of the L-cysteine dominates the bonding interactions with the Au surface. The RAS profile of Fig. 3 appears to be a characteristic profile for the thiolate/Au(110) interaction.

The RAS response of Au(110)-(1×2) and from thiolate/Au(110) surfaces can be simulated using three- and four-phase models as described earlier. The results of the simulations are shown in Fig. 4 and the values used in the simulations are listed in Table 1. The clean Au(110) surface may be simulated using equations 2 and 3 where a Lorenzian transition in $y$ simulates the transition between surface states. There is good agreement between the simulated (Fig. 4) and experimental (Fig. 3) Au(110) spectrum. For the L-cysteine saturated surface, the surface state transition is quenched

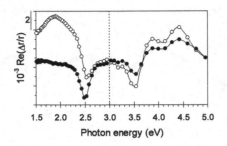

Fig. 4. Simulated RA spectra of clean Au(110) (open circles) and L-cysteine/Au(110) (filled circles). Spectra have been offset on the vertical scale to match the experimental spectra.

and a Lorenzian contribution for the thiolate molecules is not necessary - the three-phase derivative model with values listed in Table 1 is found to be sufficient to reproduce the main features of the experimental spectra (Fig. 4). This suggests that the main effect of the S-Au bonding is upon the surface-modified bands near the L point, rather than introducing any new states specific to the S-Au bond. The smaller changes in RAS for thiolate/Au (Fig. 3) relative to those observed for thiolate/Cu (Fig. 1) may be partly as a result of the greater change in surface structure for thiolate/Cu, which includes a significant surface buckling.[14]

## 5. Conclusions

RAS can yield information on bonding, orientation and ordering at the molecule/metal interface in UHV and electrochemical environments. Characteristic optical signatures are obtained from thiolate/Cu(110) and thiolate/Au(110) surfaces, and these signatures can be simulated using three- and four-phase models.

## References

1. J.F. McGilp, J. Phys.: Condens. Matter **2**, 7985 (1990).
2. D.E. Aspnes, J.P. Harbison, A.A. Studna, L.T. Florez, J. Vac. Sci. Technol. A **6**, 1327 (1988).
3. D.E. Aspnes, Mat. Sci. Eng. B **30**, 109 (1995).
4. J.F. McGilp, J. Phys.: Condens. Matter **22**, 084018 (2010).
5. P. Weightman, D.S. Martin, R.J. Cole, T. Farrell, Rep. Prog. Phys. **68**, 1251 (2005).

6. Z. Sobiesierski, D.I. Westwood, C.C. Matthai, J. Phys.: Condens. Matter **10**, 1 (1998).

7. J.-T. Zettler, Prog. Crystal Growth and Charact. **35**, 27 (1997).

8. J.F. McGilp, Prog. Surf. Sci. **49**, 1 (1995).

9. W. Richter, Phil. Trans. R. Soc. Lond. A **344**, 453 (1993).

10. D.S. Martin, P.D. Lane, G.E. Isted, R.J. Cole, N.P. Blanchard, Phys. Rev. B **82**, 075428 (2010).

11. B.G. Frederick *et al*, Phys. Rev. B **58**, 10883 (1998).

12. E. Mateo Marti, Ch. Methivier and C.M. Pradier, Langmuir **20**, 10223 (2004).

13. J.L. Domange and J. Oudar, Surf. Sci. **11**, 124 (1968).

14. A.F. Carley, P.R. Davies, R.V. Jones, K.R. Harikumar, M.W. Roberts, and C.J. Welsby, Topics in Catalysis **22**, 161 (2003).

15. J.D.E. McIntyre and D.E. Aspnes, Surf. Sci. **24**, 417 (1971).

16. L.D. Sun, M. Hohage, P. Zeppenfeld, R.E. Balderas-Navarro, and K. Hingerl, Surf. Sci. **527**, L184 (2003).

17. U. Rossow, L. Mantese, and D.E. Aspnes, J Vac. Sci. Technol. B **14**, 3070 (1996).

18. R.J. Cole, B.G. Frederick, and P. Weightman, J. Vac. Sci. Technol. A **16**, 3088 (1998).

19. D.W. Lynch and W.R. Hunter, *Handbook of Optical Constants of Solids, Vol. 1*, edited by E.D. Palik (Academic Press, New York, 1985).

20. K. Stahrenberg, Th. Herrmann, K. Wilmers, N. Esser, W. Richter, and M.J.G. Lee, Phys. Rev. B **64**, 115111 (2001).

21. S. D'Agostino, L. Chiodo, F. Della Sala, R. Cingolani, and R. Rinaldi, Phys Rev B **75**, 195444 (2007).

22. G.E. Isted, D.S. Martin, C.I. Smith, R. LeParc, R.J. Cole, P. Weightman, phys. stat. sol. (c) **2**, 4012 (2005).

23. A. Kühnle, T.R. Linderoth, B. Hammer, and F. Besenbacher, Nature **415**, 891 (2002).

24. D.S. Martin, R.J. Cole, N.P. Blanchard, G.E. Isted, D.S. Roseburgh, and P. Weightman, J. Phys.: Condens. Matter **16**, S4375 (2004).

25. A. Bowfield, C.I. Smith, M.C. Cuquerella, T. Farrell, D.G. Fernig, C. Edwards, P. Weightman, phys. stat. sol. (c) **5**, 2600 (2008).

# THE PHYSICS OF X-RAY FREE ELECTRON LASERS (X-FELS): AN ELEMENTARY APPROACH

PRIMOZ REBERNIK RIBIC

*Faculté des Sciences de Base, Ecole Polytechnique Fédérale de Lausanne (EPFL), Lausanne, Switzerland*

G. MARGARITONDO

*Faculté des Sciences de Base, Ecole Polytechnique Fédérale de Lausanne (EPFL), Lausanne, Switzerland*

Presenting the physics of X-FELs to a non-specialized audience is a challenge: not only the mathematical treatment is complex but the main properties are entangled together. We present a response to this challenge, dealing with properties such as optical gain, saturation, spectral bandwidth, electron energy spread and pulse time structure – and with the corresponding parameters. Our approach is an expansion and refinement of a previous simple treatment, improved in several ways and extended to formerly untreated aspects.

## 1. Introduction

The year 2010 was a milestone in the century-long history of x-rays [1]: four decades after the invention of the free electron lasers (FELs) [2] and 25 years after the main theoretical works [3-7] on their extension to short wavelengths, X-FELs became a reality. As a result, we were confronted by a teaching challenge: how to explain in simple terms their physics to a non-specialized audience. Surprisingly, an extensive search for suitable documents did not produce satisfactory results. The field of FELs offers a wealth of outstanding advanced publications [2-26], but we could not find a self-contained elementary approach.

In 1985 and 1988, we met [27, 28] a similar challenge for the physics of synchrotron sources. Using simple mathematical tools and basic physics arguments, we were able to explain their most fundamental properties. Can the same strategy be implemented for X-FELs?

A positive answer was given by a recent publication [29], where we demonstrated that the basic equations of X-FELs can be derived with rather simple arguments. The treatment of Ref. 29 was not complete and left out some important questions. Here, we complete the task by refining the treatment and extending the elementary approach to properties such as the time structure and to important aspects of the gain saturation mechanism.

In order to make the presentation easily readable (in harmony to its philosophy), we include here some results of Ref. 29, plus the new original parts that complement them. We hope that in this way the physics of X-FELs will become easily understandable to non-specialists.

What are the potential benefits of our effort? The answer is multi-faceted. On one hand, more than four decades of experience in synchrotron research shows that an elementary understanding of the source properties can be rather beneficial for the invention of new techniques, in particular those targeting applications outside physics.

Furthermore, in our search for an elementary X-FEL description we realized that some of the basic properties are counter-intuitive. The case was already made in Ref. 29: in essence (Fig. 1), FELs work by grouping ("microbunching") the electrons [2-26] in an accelerator with a spatial periodicity equal to the emitted wavelength. The emission occurs when the periodically microbunched electrons interact with a periodic series of magnets. As a consequence of the periodic microbunching, the emissions of electrons in different microbunches are added coherently together, leading to optical amplification. Therefore, using the FEL mechanism to produce x-rays requires microbunching of electrons with very short periodicity. At first glance, this would seem easier than obtaining a long periodicity, since the required relative electron motion is shorter. But the contrary is true: X-FELs are much more difficult to realize than long-wavelength FELs [3-7]. Why? Explaining this count-intuitive point is one of the challenges tackled in our elementary description.

As we shall see, we have reasons to claim that our approach meets these challenges. The result is, to the best of our knowledge, the (so far and by far) simplest description of the FEL and X-FEL mechanisms.

## 2. Lasing with Free Electrons: Classical Version of a Quantum Effect

All accelerator-based sources of electromagnetic waves require two basic ingredients [27, 28]: a bunch of electrons moving at a (longitudinal) speed $u \approx c$ in an accelerator and a magnet system that forces wave emission from the electrons. The most commonly used accelerators are closed-loop storage rings and linear accelerators (LINACs). The magnet system can be a single bending magnet that deflects the otherwise straight electron trajectory, or a periodic series of magnets (a "wiggler" or an "undulator") that causes the electrons to oscillate in the transverse direction.

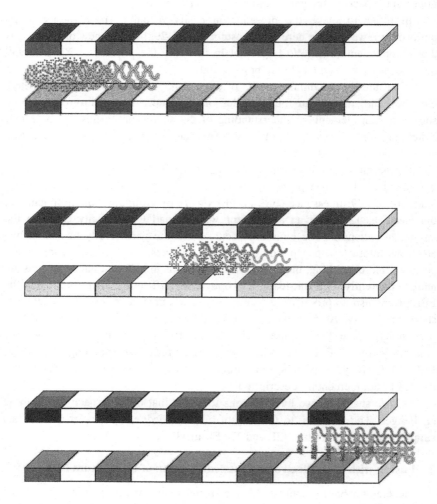

Figure 1. Electron microbunching leads to optical gain in a free electron laser. Top: a bunch of electrons traveling at nearly the speed of light enters a periodic magnet array (wiggler or undulator). In the beginning, the emission of photon waves by individual electrons is not correlated. Middle: the interaction between the electrons and the previously emitted waves starts to produce microbunches with a periodicity equal to the wavelength. The emission becomes partly correlated. Bottom: as the microbunching is complete, the emission is fully correlated.

In a synchrotron source, each electron emits waves with no correlation with respect to the other electrons [27, 28]. The total emission is therefore proportional to the total number of electrons $N$ and to $J$, the electron beam current density.

The properties of the emitted radiation are all consequences of relativity [27, 28]. Specifically, in the electron reference frame the Lorenz transformation of the periodic magnetic field gives a periodic magnetic field plus an electric field with similar periodicity that is perpendicular to it -- i.e., something similar to an electromagnetic wave. The period of both fields is determined by that of the undulator, $L$, after Lorentz contraction: in the longitudinal direction this gives $L/\gamma$, where $\gamma$ is the relativistic factor proportional to the electron energy.

Subject to the action of the periodic fields, the electron emits waves with a wavelength $L/\gamma$ in its own reference frame. In the laboratory frame, this wavelength is Doppler shifted becoming (in the longitudinal direction)

$$\lambda = \frac{L}{2\gamma^2}. \tag{1}$$

Thus, a high-energy (high-$\gamma$) accelerator *de facto* converts the macroscopic periodicity $L$ of a magnet array into the angstrom-level wavelength of the emitted x-rays.

Equation 1 is only an approximation and does not take into account the impact of the transverse electron oscillations on the longitudinal $\gamma$-factor. A more accurate version [27, 28] is provided by the widely known "undulator equation":

$$\lambda = \frac{L}{2\gamma^2}\left(1 + \frac{K^2}{2}\right), \tag{2}$$

where the undulator parameter $K$ is proportional to the (maximum) periodic $B$-field strength $B_0$ and to $L$.

Furthermore, the emission takes place not just at the wavelength of Eq. 1 (or Eq. 2) but in a wavelength band of width $\Delta\lambda$. This is in part a consequence of the duration of the emitted pulse by the electron: if the magnet array has $N_u$ periods, each electron passing through it emits a train of $N_u$ wavelengths, corresponding to a pulse time duration of $N_u\lambda/c$. The Fourier transform of this duration yields the following bandwidth:

$$\frac{\Delta\lambda}{\lambda} = \frac{1}{N_u}; \tag{3}$$

we shall see, however, that other factors contribute to the real spectral bandwidth of a FEL, so that Eq. 3 gives only a minimum value.

The above emission properties – in particular the wavelength – are still present when the emissions of individual electrons become correlated, in a FEL or X-FEL [3-26]. However, the emissions of individual electrons combine together in a different way: grasping this point is the key to understanding the lasing mechanisms.

Assume that, as mentioned above and for reasons to be clarified later, the electrons are not homogeneously distributed in the bunch but, as already mentioned, periodically distributed in microbunches – see Fig. 1 [29]. Furthermore, assume that the microbunch periodicity equals the wavelength. Then the cause of the correlated combination of the individual electron emissions is immediately clear: all electrons in a microbunch oscillate together and the waves emitted by two different microbunches are in phase. The total emitted wave is a combination not of the intensities of the individual electron emissions but of their amplitudes: its intensity is proportional to the square of the total amplitude and therefore to $N^2$ and $J^2$ rather than to $N$ and $J$.

A key question, however, remains: why do the electrons become microbunched? The answer is subtle: the electrons do not travel alone but together with their previously emitted waves [2-26]. In a classical picture [29], the magnetic field of the waves and the transverse (oscillation) velocity of the electrons create a Lorentz force in the longitudinal direction. This is the force that causes microbunching: as we shall see, however, the detailed mechanism is not trivial.

The fact that microbunching is triggered by previously emitted waves has a fundamental consequence: the additional emission intensity is proportional [29] to the pre-existing emission intensity: $dI/dt = (u/L_G)I$ (where we wrote the proportionality constant in terms of the longitudinal electron speed and the "gain length" parameter $L_G$). This gives an exponential intensity increase – or gain – in the longitudinal direction (see Fig. 2):

$$I = I_o \exp\left(\frac{ut}{L_G}\right) = I_o \exp\left(\frac{x}{L_G}\right),$$ (4)

where $x = ut$ is the longitudinal coordinate.

Equation 4 shows how microbunching produces optical amplification. This is similar to what happens in a normal laser, for which, however, the optical amplification is a quantum effect related to stimulated emission. In a sense, the

FEL and X-FEL mechanisms are classical counterparts of the quantum optical amplification in lasers.

We would actually like to propose a more imaginative picture. Microbunching is equivalent to the creation of nanostructures within the electron bunch. Therefore, it is a form of nanofabrication applied not to a solid system but to electrons moving in vacuum. This elegant albeit short-lived "nanofabrication in vacuum" causes optical amplification in FELs and X-FELs.

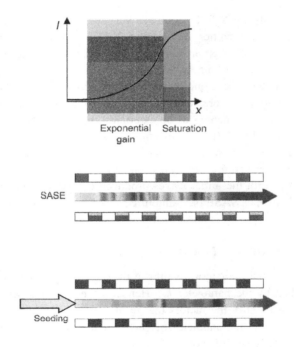

Figure 2. Top: the optical gain produces an exponential increase of the photon beam along the undulator path, until saturation occurs as discussed in the text. Middle and bottom: the initial wave could be either produced by the initial emission of the electrons in the undulator (SASE = self amplified spontaneous emission) or injected from outside (seeding).

## 3. Optical Amplification is Here: So What?

Having achieved optical amplification, can we assume that it will continue indefinitely? The answer is no, for several reasons. Of course, amplification can only occur over the length of the magnet array. Quite often, however, it saturates (see Fig. 2) [2-26, 29] before the end of the array, at a distance $L_S$. One could imagine that gain saturation simply occurs when all electrons are microbunched. We shall see, however, that the actual mechanism is more complicated.

If the magnet array is too short with respect to the "saturation length" $L_S$, then the potential gain is not fully exploited. Conventional laser technology has a remedy for that: the use of an external optical cavity (two mirrors) to expand the path available for amplification. The remedy is also used for infrared-emitting FELs, but it does not work for X-FELs: efficient (normal incidence) mirrors do not exist for x-rays.

Without a cavity, an X-FEL requires a very long undulator. This is a partial explanation of the difficulties in building X-FELs. We shall see later that additional problems exist: as already mentioned, microbunching is much more difficult for x-rays than for long wavelengths.

We shall conclude these qualitative arguments by noting one last important point. In the previous picture, nothing prevents (Fig. 2) the initial wave that triggers electron microbunching to be externally injected rather than being the initial electron emission [29]. This injection or "seeding" can be a very effective way to operate a FEL, for reasons that will be discussed later and are primarily related to the time structure of the emission. When no seeding is used, the FEL mechanism starts with the initial waves produced by spontaneous emission and the process is known as SASE (self-amplified spontaneous emission) – see Fig. 2 [3-7].

## 4. A Simple Description of the Optical Amplification

Equation 4 shows that the major features of the optical amplification mechanism can be described by the parameter $L_G$. Hence, a theoretical description of this mechanism must identify the factors that influence this parameter. As we shall see, the main ones are the undulator magnetic field strength and period, as well as the electron relativistic $\gamma$-factor and current density.

Optical amplification means the transfer of energy from each electron to the previously emitted wave. The rate of this transfer is proportional to the product $E_W v_T$, where $E_W$ is the magnitude of the wave electric field (proportional to the square root of the wave intensity):

$$E_W = \text{constant} \times \sqrt{I} \times \cos\left[2\pi\left(\frac{x}{\lambda} - \frac{ct}{\lambda}\right) + \phi\right] =$$
$$= \text{constant} \times \sqrt{I} \times \cos\left[2\pi\left(\frac{ut}{\lambda} - \frac{ct}{\lambda}\right) + \phi\right] \tag{5}$$

and $v_T$ is the electron transverse velocity. In turn, $v_T$ is caused by the Lorentz force of magnitude $-euB$, created by the undulator periodic magnetic field, whose amplitude we can write as:

$$B = B_o \sin\left(\frac{2\pi u t}{L}\right). \tag{6}$$

The effects of this Lorentz force can be derived using the standard transverse relativistic dynamics equation:

$$\gamma m_o \frac{dv_T}{dt} = \text{transverse force} = -euB_o \sin\left(\frac{2\pi u t}{L}\right). \tag{7}$$

The solution of this equation for small transverse velocities (and $\gamma \approx$ constant) is:

$$v_T = \left(\frac{euB_o}{\gamma m_0}\right)\left(\frac{L}{2\pi u}\right)\cos\left(\frac{2\pi u t}{L}\right). \tag{8}$$

The electron $\rightarrow$ wave energy transfer rate is, therefore:

$$E_w v_T = \text{constant} \times \sqrt{I} \times \cos\left[2\pi\left(\frac{ut}{\lambda} - \frac{ct}{\lambda}\right) + \phi\right] \times \left(\frac{euB_o}{\gamma m_0}\right)\left(\frac{L}{2\pi u}\right)\cos\left(\frac{2\pi u t}{L}\right). \tag{9}$$

Equation 9 describes the effects of a single electron. We must now evaluate the collective effects of the electrons in the bunch. If all electrons were arranged in microbunches, then the total transfer to the wave would be simply proportional to $N^2$, as discussed above. At the beginning of the undulator, however, the electrons are not microbunched: their spatial re-arrangement occurs afterwards, as they interact with previously emitted waves.

The longitudinal force that produces the electron microbunching is the Lorentz force, of magnitude $-eB_W v_T$, created by the magnetic field of the previously emitted wave, of magnitude $B_W$, and by the electron transverse velocity, $v_T$. Using Eq. 8 and by analogy with Eq. 5, we can write it as:

longitudinal microbunching force = $-eB_W v_T$ =

$$= \text{constant} \times \left(\frac{euB_o}{\gamma m_0}\right)\left(\frac{L}{2\pi u}\right)\cos\left(\frac{2\pi u t}{L}\right) \times \sqrt{I} \times \cos\left[2\pi\left(\frac{ut}{\lambda} - \frac{ct}{\lambda}\right) + \phi\right]. \tag{10}$$

The effect of this force is a small longitudinal shift $\delta x$ of the electron position superimposed to the main motion with speed $u$. The shift can be derived using the relativistic equation for longitudinal dynamics:

$$\gamma^3 m_o \frac{d^2(\delta x)}{dt^2} = \text{longitudinal force} = -eB_W v_T =$$

$$= \text{constant} \times \left(\frac{euB_o}{\gamma m_o}\right)\left(\frac{L}{2\pi u}\right)\cos\left(\frac{2\pi ut}{L}\right) \times \sqrt{I} \times \cos\left[2\pi\left(\frac{ut}{\lambda} - \frac{ct}{\lambda}\right) + \phi\right]. \quad (11)$$

The solution of this equation seems quite complicated even for small shifts and $\gamma \approx$ constant. We can simplify our mathematical task by first focusing our attention on the product:

$$\cos\left(\frac{2\pi ut}{L}\right)\cos\left[2\pi\left(\frac{ut}{\lambda} - \frac{ct}{\lambda}\right) + \phi\right]$$

on the right-hand side of Eq. 11. Elementary trigonometry says that $2\cos(\alpha)\cos(\beta) = \cos(\alpha+\beta) + \cos(\alpha-\beta)$, so this product is proportional to:

$$\cos\left[2\pi\left(\frac{ut}{L} + \frac{ut}{\lambda} - \frac{ct}{\lambda}\right) + \phi\right] + \cos\left[2\pi\left(\frac{ut}{L} - \frac{ut}{\lambda} + \frac{ct}{\lambda}\right) - \phi\right]. \quad (12)$$

The argument of the second cosine term can be written as:

$$2\pi\left(\frac{ut}{L} - \frac{ut}{\lambda} + \frac{ct}{\lambda}\right) - \phi = 2\pi\left(\frac{1}{L} - \frac{1}{\lambda} + \frac{c}{u\lambda}\right)ut - \phi = 2\pi\left[\frac{1}{L} + \frac{c}{u}\left(1 - \frac{u}{c}\right)\frac{1}{\lambda}\right]ut - \phi; \quad (13)$$

note that $\quad \frac{1}{\gamma^2} = 1 - \frac{u^2}{c^2} = \frac{c^2 - u^2}{c^2} = \frac{(c+u)(c-u)}{c^2} \approx \frac{2(c-u)}{c} = 2\left(1 - \frac{u}{c}\right), \quad (14)$

thus we can re-write the term in (13) as:

$$\approx 2\pi\left[\frac{1}{L} + \frac{1}{2\gamma^2}\frac{1}{\lambda}\right]ut - \phi,$$

and finally, using Eq. 1, as $\approx 4\pi ut/L - \phi$. This corresponds to a very fast oscillation with a zero average whose effects can be neglected.

The argument of the first cosine term of Eq. 12 can be evaluated in a similar way; using again Eq. 1, we obtain:

$$\approx 2\pi\left[\frac{1}{L} - \frac{1}{2\gamma^2}\frac{1}{\lambda}\right]ut + \phi \approx \phi, \quad (15)$$

i.e., a constant argument that gives a constant cosine. Equation 11 can thus be drastically simplified to:

$$\gamma^3 m_o \frac{d^2(\delta x)}{dt^2} = \text{constant} \times \left(\frac{euB_o}{\gamma m_o}\right)\left(\frac{L}{2\pi u}\right)\sqrt{I}. \tag{16}$$

The solution of this equation is quite straightforward if we use the gain equation (4) as an empirical fact based on recent experimental data [1]; then Eq. 16 can be written as:

$$\gamma^3 m_o \frac{d^2(\delta x)}{dt^2} = \text{constant} \times \left(\frac{eB_oL}{2\pi\gamma m_o}\right)\sqrt{I_0}\,\exp\left(\frac{ut}{2L_G}\right),$$

whose solution is simply, assuming a negligible wave intensity for zero $\delta x$, i.e., when microbunching starts:

$$\delta x = \text{constant} \times \left(\frac{eB_oL}{2\pi\gamma^4 m_o^2}\right)L_G^2\sqrt{I_0}\,\exp\left(\frac{ut}{2L_G}\right) = \text{constant} \times \left(\frac{eB_oLL_G^2}{2\pi\gamma^4 m_o^2}\right)\sqrt{I}. \tag{17}$$

The physical meaning of Eq. 17 is clear: in addition to moving with speed $u$, the electrons start to migrate towards microbunches with the additional small shift $\delta x$. Complete microbunching is achieved when all electrons are in very narrow slabs –microbunches – separated by a distance equivalent to the wavelength $\lambda$.

Before complete microbunching, only a portion of the electrons emit in a correlated fashion. We can simply assume that this portion is proportional, in first approximation, to the ratio between the actual shift $\delta x$, and the total microbunching shift $\lambda$. This also corresponds to the fraction of the electron density $J$ that participates in the correlated emission.

The total correlated emission is thus proportional to the individual correlated emission, Eq. 9, multiplied by the factor $(\delta x/\lambda)J$. Note that we can apply the same argument as we did for Eq. 11 to the product of the two cosines in Eq. 9. Putting everything together and using Eq 17 for $\delta x$, we obtain:

$$\frac{dI}{dt} = \text{total correlated emission rate} =$$

$$= \text{constant} \times \left(\frac{eB_oLL_G^2}{2\pi\gamma^4 m_o^2}\right)\sqrt{I}\,\frac{1}{\lambda}J\sqrt{I}\left(\frac{euB_o}{\gamma m_o}\right)\left(\frac{L}{2\pi u}\right) =$$

$$= \text{constant} \times \left(\frac{e^2 B_o^2 L^2 L_G^2}{4\pi^2\gamma^5 m_o^3}\right)\frac{J}{\lambda}I,$$

which gives, using Eq. 1:

$$\frac{dI}{dt} = \text{constant} \times \left( \frac{e^2 B_o{}^2 L L_G{}^2 J}{2\pi^2 \gamma^3 m_o^3} \right) I.$$

This is the key result as far as optical gain is concerned. Indeed, this equation has the required form, $dI/dt = (u/L_G)I$, to obtain the gain law, Eq. 4, if $u/L_G$ is identified with the coefficient in front of $I$; grouping the universal parameters $e$, $m_o$ and $\pi$ into one constant, this gives:

$$\frac{u}{L_G} = \text{constant} \times \left( \frac{B_o{}^2 L L_G{}^2 J}{\gamma^3} \right) \tag{18}$$

and finally:

$$L_G = \text{constant} \times J^{-1/3} B_o{}^{-2/3} L^{1/3} \gamma, \tag{19}$$

one of the main results of rigorous FEL theories, which identifies the key parameters playing a role in the gain length [3, 13].

Quite often, the same result is presented using the so-called "FEL parameter" [3] defined as:

$$\rho = \frac{L}{4\pi\sqrt{3}L_G}, \tag{20}$$

so that Eq. 19 implies:

$$\rho = \text{constant} \times J^{1/3} B_o{}^{2/3} L^{4/3} \gamma^{-1}, \tag{21}$$

in agreement with its rigorous theoretical definition [3].

## 5. Theory Refinement

In the above derivation, we skipped some details that merit further attention. First, we did not take fully into account the energy transfer from the electrons to the wave. One could imagine that microbunching cannot go on because the speed of the electrons changes as a result of this energy exchange. However, we will demonstrate here that, in first approximation, this effect contributes to the formation of microbunches in a similar way as the Lorentz force caused by the wave magnetic field.

As we have seen, an electron travels with a transverse oscillatory motion. Suppose that at a given time $t$ the electron velocity vector $v$ forms an angle $\theta$ with the undulator axis as shown in Fig. 3. In this case, the wave magnetic field vector $\boldsymbol{B}_W$, which points out of Fig. 3, decreases the longitudinal velocity $u$ of the electron: this leads to microbunching as described previously.

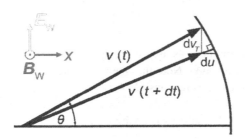

Figure 3. Due to the undulator magnetic field the electron undergoes oscillatory motion. Assume that at a time $t$ the electron velocity makes an angle $\theta$ with the undulator axis. The wave magnetic field results in a force that reduces the longitudinal velocity of the electron, which eventually leads to microbunching. This effect is accompanied by the energy transfer from the electron to the wave. The wave electric field decreases the transverse electron velocity and as a result the magnitude of the electron velocity $v$ is also decreased. This also corresponds to a decrease of the average longitudinal velocity of the electron, which contributes to microbunching.

The energy transfer electron $\rightarrow$ wave is due, as we have seen, to the force on the electron caused by the wave electric field $\boldsymbol{E}_W$. The wave electric field points in the same direction as the transverse velocity. The force due to this field is $-e\boldsymbol{E}_W$ and points in the opposite direction of the transverse velocity and thus decreases it. As a result, the magnitude $v$ of the electron velocity $\boldsymbol{v}$ also decreases. We can evaluate its time derivative by using the relativistic equation for the total electron energy $W$:

$$\frac{dW}{dt} = -e\boldsymbol{E} \cdot \boldsymbol{v}, \tag{22}$$

where $W = \gamma \square m_o c^2$. Since the electric field is parallel to the transverse velocity, we can write Eq. 22 as:

$$m_o c^2 \frac{d\gamma}{dt} = -eE_W v_T. \tag{23}$$

Now, since:

$$\frac{d\gamma}{dt} = \frac{1}{c^2} \gamma^3 v \frac{dv}{dt} \tag{24}$$

and $v \approx u \approx c$, we obtain from Eq. 23:

$$\gamma^3 m_0 \frac{du}{dt} \approx -e \frac{E_W}{c} v_T. \tag{25}$$

For an electromagnetic wave, $E_W = cB_W$, thus Eq. 25 can be written as:

$$\gamma^3 m_0 \frac{du}{dt} \approx -eB_W v_T. \tag{26}$$

This equation implies, as expected, that the magnitude of the electron velocity decreases with time. Note that the right-hand term of Eq. 26 is equal to the longitudinal force that is responsible for microbunching (Eq. 11).

Are these two facts — velocity magnitude decrease and microbunching — related to each other? The answer is positive since:

$$u = \text{constant} + \frac{d(\delta x)}{dt}, \tag{27}$$

and therefore:

$$\frac{du}{dt} = \frac{d^2(\delta x)}{dt^2}. \tag{28}$$

Thus, Eq. 26 is approximately equivalent to:

$$\gamma^3 m_0 \frac{d^2(\delta x)}{dt^2} \approx -eB_W v_T, \tag{29}$$

and this is nothing but Eq. 11, which explains microbunching.

Thus, at this level of approximation, the effect of the electron → wave energy transfer is identical to the effect of the wave magnetic field on the electron, which is responsible for microbunching. Emission of electromagnetic radiation from the electrons does not interfere with the formation of microbunches and with the resulting optical gain; on the contrary, it is an aspect of the microbunching mechanism. We shall see shortly, however, that this conclusion must be partially revised when the effects of the energy transfer on the phase of the electron transverse velocity are taken into account.

## 6. Phase Effects: Gain Saturation

Phase effects are in fact the key to understanding why the optical gain does not go on indefinitely but saturates [3, 13]. We can start our discussion of this key point from the energy transfer rate, Eq. 9. This equation includes precisely the same product of cosines (or "phase terms"),

$$\cos\left(\frac{2\pi ut}{L}\right)\cos\left[2\pi\left(\frac{ut}{\lambda}-\frac{ct}{\lambda}\right)+\phi\right],$$

that we found in Eq. 11 while treating microbunching. We can thus use again our discussion of this product, in particular Eq. 15, to conclude that the only part that matters is the term:

$$\cos\left[2\pi\left(\frac{ut}{L}+\frac{ut}{\lambda}-\frac{ct}{\lambda}\right)+\phi\right], \tag{30}$$

which according to Eq. 15 is approximately equal to $\cos(\phi)$ = constant. Thus, the energy transfer rate stays constant with time at this level of approximation.

What is the physics underlying this result? The electron → wave energy transfer requires the electric field of the wave to do negative work. This occurs if the corresponding force points in the opposite direction of the (transverse) velocity of the electron.

Assume that at a certain time this condition is met and there is indeed electron → wave energy transfer. Will this condition continue to be present at subsequent times? The answer is positive as long as Eq. 15 is valid, since the energy transfer is proportional to $\cos(\phi)$ and therefore constant with respect to $t$.

The microbunching mechanism is also linked to the validity of Eq. 15, required to transform Eq. 11 into Eq. 16. Otherwise, the microbunching force would not be constant: if it points in the direction favorable for microbunching at a given time, it could go against it later!

In conclusion, both for microbunching and for the energy transfer by individual electrons, Eq. 15 plays a key role. But is it always applicable? Let us see what are the conditions for its validity.

In essence, Eq. 15 is a consequence of Eq. 1 that links the wavelength, the undulator period and the electron $\gamma$-factor, which in turn is related to the electron speed. As long as Eq. 1 is valid, the time-dependent term in Eq. 15 is zero and both the energy transfer rate and the microbunching force stay constant with respect to time.

This is quite a remarkable result: note, in particular, that if the electron traveled exactly at the speed of light as the wave, we would have $1/\gamma^2 = 0$ and Eq. 15 would *not* be valid. Constant energy transfer and microbunching are the result of the small difference between the electron speed $u$ and $c$: the magnitude of this difference is precisely the right one to guarantee the validity of Eq. 15.

All these conclusions, however, are based on the same assumption: the $\gamma$ factor of the electron is constant, i.e., the electron energy is constant. This assumption, however, is in conflict with the energy transfer electron → wave that decreases the electron speed, the electron energy and the $\gamma$-factor. If Eq. 15 is valid at a certain time, it will not be strictly valid later because of the electron energy loss that changes the $\gamma$-factor to $\gamma - \delta\gamma$.

This impacts both the energy transfer rate and the microbunching (Eqs. 9 and 11): they are no longer constant but cosine functions, i.e., oscillating functions of time of the approximate form (according to Eq. 12 and 15 and neglecting once again the fast-oscillating term):

$$\cos\left\{2\pi\left[\frac{1}{L} - \frac{1}{2(\gamma-\delta\gamma)^2}\frac{1}{\lambda}\right]ut + \phi\right\}. \tag{31}$$

For $\delta\gamma \ll \gamma$, $1/(\gamma \;\square\square\; \delta\gamma)^2 \approx (\gamma + \square\square\; \delta\gamma)/\gamma^3$; the cosine in (31) is thus approximately:

$$\cos\left[2\pi\left(\frac{\delta\gamma}{\gamma^3\lambda}\right)ut + \phi\right], \tag{32}$$

whose (angular) oscillation frequency is $2\pi\delta\gamma u/(\gamma^3\lambda)$.

Since $W = \gamma\;\square\;m_o c^2$, we have $\delta\gamma/\gamma = \delta W/W$ (where $\delta W$ is the electron energy loss), and this oscillation frequency can be written as $2\pi(\delta W/W)[u/(\gamma^2\lambda)]$, or, using once again Eq. 1:

$$\text{angular oscillation frequency} \approx 4\pi\left(\frac{\delta W}{W}\right)\left(\frac{u}{L}\right). \tag{33}$$

What are the effects of the oscillation of the energy transfer rate? Initially, the energy loss and therefore $\delta\gamma$ are small, thus the frequency is low. If the electron travels over the gain length in a time $L_G/u$ much shorter than the oscillation period $(W/\delta W)[L/(2u)]$, the oscillation does not prevent steady gain to occur.

However, as the energy loss increases the oscillation period decreases and the effects of the oscillation are no longer negligible over the time $L_G/u$: eventually, instead of steady gain there is gain saturation. This saturation condition can be accurately written as [4]:

$$\text{angular oscillation frequency} \approx \text{energy transfer rate} = u/L_G,$$

which gives:

$$\frac{L}{L_G} \approx 4\pi\frac{\delta W}{W}. \tag{34}$$

Equation 34 reveals that the gain length also corresponds to the fraction of the electron energy that is transferred to the wave before saturation takes place. Note that typically $L \ll L_G$, therefore the relative energy transfer $\delta W/W$ is small. The result of Eq. 34 can be expressed in terms of the FEL parameter [3]: combined with Eq. 20, Eq. 34 can be written as:

$$\frac{\delta W}{W} \approx \sqrt{3}\rho, \tag{35}$$

implying that the FEL parameter measures how effectively the energy transfers from the electrons to the wave.

How far do the electrons travel after entering the magnet array and until saturation occurs? This distance is by definition the saturation length $L_S$ and can be roughly evaluated in the following way. We have seen that the energy transfer is an oscillating function, going from the electron $\rightarrow$ wave direction to wave $\rightarrow$ electron, then to electron $\rightarrow$ wave etc., with increasing (angular) frequency that reaches the value $u/L_G$ at saturation. We can approximate the saturation length $L_S$ as the distance which the electron travels during the time for which the transfer remains in the electron $\rightarrow$ wave direction, i.e., during one-quarter to one-half of the first oscillation.

If the angular frequency was constant and equal to $u/L_G$, then a one-quarter period would be $\pi L_G/(2u)$ and $L_S$ would be in the order of $[\pi L_G/(2u)]u = (\pi/2)L_G$. The frequency, however, is not constant but increases from zero reaching $\approx u/L_G$ at saturation. Assume, as a rough approximation, that the frequency increases linearly from zero to $u/L_G$ over the time $T_S$ required to reach saturation:

$$\text{angular frequency} = \left(\frac{u}{L_G}\right)\left(\frac{t}{T_s}\right); \tag{36}$$

then, in a time $dt$ the phase of the cosine function in Eq. 32 would change by:

$$\left(\frac{u}{L_G}\right)\left(\frac{t}{T_s}\right)dt;$$

after integration, this gives a cosine phase of:

$$\frac{ut^2}{2L_G T_s} + \phi. \tag{37}$$

If the starting phase is zero (maximum energy transfer), i.e., if $\phi = 0$, then the energy transfer direction is reversed at a phase of $\pi/2$, so that Eq. 37 gives:

$$\frac{uT_S{}^2}{2L_G T_S} = \frac{uT_S}{2L_G} \approx \frac{\pi}{2},$$

corresponding to $L_S = uT_S \approx \pi L_G$. These arguments, however approximated, lead us to the realization that the saturation length is linearly related to the gain length:

$$L_S = \text{constant} \times L_G. \tag{38}$$

Equations 34, 35 and 38 justify the main properties of the saturation process. They are qualitatively consistent with the results of theoretical models [3, 4, 13] much more sophisticated than ours. Even quantitatively they are not too far from such results; their accurate versions are indeed [3, 4, 8, 9, 13]:

$$\frac{L}{L_G} \approx 4\pi\sqrt{3}\,\frac{\delta W}{W}, \tag{39}$$

$$\frac{\delta W}{W} \approx \rho, \tag{40}$$

$$L_S \approx 4\pi\sqrt{3}\,L_G \approx 22\,L_G. \tag{41}$$

## 7. Characteristics of the FEL Pulse

The emitted pulse characteristics are important factors that make FEL sources very attractive. Such characteristics - in particular, the time structure, the spectral bandwidth, the peak power and longitudinal coherence - are a result of complex interactions between the emitted electromagnetic wave and the individual electrons in the bunch. For a complete and accurate description of these properties, the Maxwell equations must in principle be solved together with the equations of motion [13, 25]. However, we will now show that it is possible to understand some of the basic pulse parameters using simple physical arguments.

### 7.1. Pulse Duration

Consider an electron bunch of length $L_B$ that enters an undulator at a time $t = 0$ as shown in the top panel of Fig. 4. The first emitted photons reach the detector at a time $t_1 = (N_u L + D)/c$, where $D$ is the distance of the detector from the undulator exit. The electron bunch emits electromagnetic radiation as it moves through the undulator. The last photons are emitted at $t = (N_u L + L_B)/u$ when the electron bunch exits the undulator (Fig. 4, bottom panel). These photons arrive at the detector at the time $t_2 = (N_u L + L_B)/u + D/c$. The total duration of the laser pulse $\Delta t$ is thus:

$$\Delta t = t_2 - t_1 = \frac{N_u L + L_B}{u} + \frac{D}{c} - \frac{N_u L + D}{c} = \frac{L_B}{u} + N_u L \left( \frac{1}{u} - \frac{1}{c} \right) =$$

$$= \frac{L_B}{u} + N_u L \frac{c - u}{uc}; \qquad (42)$$

using Eqs. 1 and Eq. 14, we can write Eq. 42 as:

$$\Delta t = \frac{L_B}{c} + \frac{N_u \lambda}{c}. \qquad (43)$$

In the one-electron case, we can neglect the bunch length $L_B$ : the emitted FEL pulse duration is simply determined by the "electron pulse duration", $N_u \lambda / c$ – the time taken by the wave train $N_u \lambda$ to pass through a given point. However, Eq. 43 shows that the actual pulse duration can be tuned to values different from $N_u \lambda / c$ by changing $L_B$.

For X-FELs, the typical emitted pulse duration is on the order of $10^2$ fs. By using bunch compressors to decrease the electron bunch length, the pulse duration can be reduced to 10 fs or less [1].

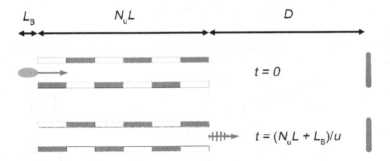

Figure 4. Estimation of the laser pulse duration: the electrons in the bunch start emitting electromagnetic radiation when the bunch front enters the undulator (top) and the last photons are emitted when the electron bunch end exits the undulator (bottom). The total duration of the laser pulse can be estimated from the total length of the undulator and the electron bunch length.

These simple considerations explain, at least in part, why the FEL pulses are very short in time. However, they do not provide information about the detailed time structure of each pulse, which can be very complicated. Specifically, the time structure of a SASE X-FEL pulse consists of a number of "spikes" caused by the electron density fluctuations (shot noise) in the electron bunch when it enters the undulator. Because of the Fourier transform properties, the spikes in the time domain correspond to a "spiky" photon frequency (or wavelength) spectrum. For a proper description of these phenomena, sophisticated models must be used [25].

## 7.2. *Spectral bandwidth*

In a seeded (non-SASE) FEL, the spectrum has a narrow bandwidth $\Delta v$ around the seed photon frequency. This bandwidth is related to the emitted pulse duration by the Fourier transform properties that give $\Delta v \square \square \approx 1/\Delta t$ . For an individual electron, $\Delta t$ is ideally determined by the transit time in the undulator, leading to Eq. 3.

We shall now expand our analysis of this point. In a SASE FEL, a range of wavelengths around the resonance wavelength of Eq. 1 (or more accurately of Eq. 2) is amplified as a result of the FEL startup process from shot noise. We can describe this phenomenon and its impact on the spectral bandwidth in simple terms. Assume indeed that for a wavelength $\lambda$ the electromagnetic wave is in resonance (Eq. 1) and has the right phase to be amplified. The electric field of this wave (see Eq. 5) is proportional to:

$$\cos\left[2\pi\left(\frac{x}{\lambda}-\frac{ct}{\lambda}\right)+\phi\right]=\cos\left[2\pi\left(\frac{ut}{\lambda}-\frac{ct}{\lambda}\right)+\phi\right].$$

Suppose now that the wavelength is not $\lambda$ but $\lambda + \Delta\lambda$, with $\Delta\lambda \ll \lambda$. In this case, the wave is proportional to:

$$\cos\left[2\pi\left(\frac{ut}{\lambda+\Delta\lambda}-\frac{ct}{\lambda+\Delta\lambda}\right)+\phi\right]\approx\cos\left[2\pi\left(\frac{ut}{\lambda}-\frac{ct}{\lambda}\right)\left(1-\frac{\Delta\lambda}{\lambda}\right)+\phi\right]. \quad (44)$$

The phase difference with respect to the wave in resonance is:

$$\approx 2\pi\frac{(u-c)t}{\lambda^2}\Delta\lambda=-2\pi c\left(1-\frac{u}{c}\right)\frac{t\Delta\lambda}{\lambda^2}=-2\pi c\frac{\left(1-\frac{u}{c}\right)\left(1+\frac{u}{c}\right)}{\left(1+\frac{u}{c}\right)}\frac{t\Delta\lambda}{\lambda^2}\approx$$

$$\approx -2\pi c\frac{1}{2\gamma^2}\frac{t\Delta\lambda}{\lambda^2}, \quad (45)$$

which, using Eq. 1, can be written as:

$$-2\pi c\frac{t\Delta\lambda}{L\lambda}. \quad (46)$$

When the magnitude of this phase difference becomes too big, the energy transfer conditions are no longer satisfied and the wave with a wavelength $\lambda + \Delta\lambda$ is not amplified. This condition can be approximately written as:

$$-2\pi c \frac{t\Delta\lambda}{L\lambda} = \pi. \tag{47}$$

From Eq. 47, we can estimate the bandwidth $\Delta\lambda$ at saturation ($t \approx L_S/u \approx L_S/c$):

$$\frac{\Delta\lambda}{\lambda} \approx \frac{1}{2}\frac{L}{L_S}. \tag{48}$$

By taking into account Eqs. 20 and 38, the above equation can be written as:

$$\frac{\Delta\lambda}{\lambda} \approx \frac{\sqrt{3}}{2}\rho, \tag{49}$$

a result that links the spectral bandwidth and the FEL parameter – and through it the gain and saturation lengths – and is quite consistent with more elaborate treatments of the FEL theory, that give:

$$\frac{\Delta\lambda}{\lambda} \approx \rho. \tag{50}$$

Note that a similar result is obtained when analyzing the effects of the electron energy spread. Indeed, these can be roughly estimated by differentiating the logarithm of Eq. 1, obtaining:

$$\frac{\Delta\lambda}{\lambda} = 2\frac{\Delta\gamma}{\gamma} = 2\frac{\Delta W}{W} \tag{51}$$

and discovering that the spectral bandwidth is directly linked to the energy spread.

### 7.3. *Radiated power*

The average radiated power for a single electron moving through an undulator is (see, e.g., Ref. 24):

$$\langle P \rangle_1 = \frac{ce^2\pi K^2\gamma^2}{3\varepsilon_0 L^2}. \tag{52}$$

If the emission from electrons in the bunch is uncorrelated then the total emitted power is just the sum of individual power contributions: $\langle P \rangle_{tot} = N\langle P \rangle_1$. We have already seen that if all the electrons in the bunch emit in a fully correlated way, then the electric fields produced by the individual electrons add together and the total emitted power, which is proportional to the square of the total electric field, becomes:

$$<P>_{tot} = N^2 <P>_1. \tag{53}$$

For example, for the Vacuum Ultraviolet FEL (VUV-FEL) at DESY in Hamburg, operating at $\lambda = 32$ nm, the power $<P>_1$ evaluated with Eq. 52 is $\approx 1.33 \times 10^{-9}$ W (undulator and beam parameters were taken from Ref. 30). From the total charge of the electron bunch, 1 nC, the estimated number of electrons is $N \approx 6.25 \times 10^9$. For uncorrelated emission, this results in a total radiated power of $\approx 8.3$ W. For the fully correlated case, the total emitted power given by Eq. 53 would be 52 GW : almost 10 orders of magnitude higher !

The average experimental value for the radiated power obtained from Ref. 30 is actually lower, 0.4 GW. This means in particular that only a fraction of electrons in the bunch emit in a correlated way. If we neglect other factors that can lower the emission, this fraction can be estimated to be $\approx 10\%$.

## 8. Did We Keep Our Promises?

We trust we did, at least in part, and the first benchmark is the solution of the apparent paradox: why is short-periodicity microbunching in an X-FEL much more difficult to achieve than long-period microbunching? The answer can be found in Eq. 11: an X-FEL requires high-energy electrons with a large $\gamma$-factor. Thus, the longitudinal relativistic mass $\gamma^3 m_o$ becomes exceedingly large, making the electrons very difficult to move. Whereas short-period microbunching requires a shorter relative electron motion $\delta x$ than large-period microbunching, the electrons that must be moved are much, much "heavier" – and this factor more than offsets the shortness of $\delta x$.

Actually, high-energy electrons are also heavy in the transverse direction, the relativistic transverse mass being $\gamma m_o$. This further complicates the task of building an X-FEL, since (Eq. 7) it makes it more difficult to get a sufficiently large transverse velocity with the undulator magnetic field.

Besides the initial challenge of explaining the above puzzle, our treatment achieved other objectives. In particular, by discussing in an elementary way individual properties such as gain, saturation, bandwidth and energy spread, it revealed a number of links between them. In fact, the basic aspects of the FEL operation appear all related to each other, making FEL physics both intriguing and not trivial to understand.

Our approach should not be used *in lieu* of full theories that can handle many more phenomena and are much more rigorous. We believe, however, that with our treatment the physics underlying the phenomena emerges in a correct albeit approximated way: the main challenge that we hope to have met.

## Acknowledgments

This work was supported by the Fonds National Suisse pour la Recherche Scientifique and by the Center for Biomolecular Imaging (CIBM), in turn supported by the Louis-Jeantet and Leenaards foundations.

## References

1. Emma, P., Akre, R., Arthur, J., Bionta, R., Bostedt, C., Bozek, J., Brachmann, A., Buscksbaum, P., Coffee, R., Decker, F.-J., Ding, Y., Dowell, D., Edstrom, S., Fisher, A., Frisch, J., Gilevich, S., Hastings, J., Hays, G., Hering, Ph., Huang, Z., Iverson, R., Loos, H., Messerchmidt, M., Miahnahri, A., Moeller, S., Nuhn, H.-D., Pile, G., Ratner, D., Rzepiela, J., Schultz, D., Smith, T., Stefan, P., Tompkins, H., Turner, J., Welch, J., White, W., Wu, J., Yocky, G. & Galayda, J. (2010). Nature Photonics **4**, 641-647.
2. Madey, J. (1971). J. Appl. Phys. **42**, 1906-1913.
3. Bonifacio, R., Pellegrini C. & Narducci L. M. (1984). Opt. Commun. **50**, 373-378.
4. Murphy, J. B. & Pellegrini, C. (1985). *Introduction to the physics of free-electron lasers*, in Laser Handbook, edited by W. B. Colson, C. Pellegrini & A. Renieri. Amsterdam: North-Holland.
5. Bonifacio, R., De Salvo, L., Pierini, P., Piovella, N. & Pellegrini C. (1994). Phys. Rev. Lett. **73**, 70-73.
6. Pellegrini, C. (2000). Nucl. Instrum. Methods Phys. Res. **A 445**, 124-127.
7. Bonifacio, R. & Casagrande, S. (1985). J. Opt. Soc. Am. **B2**, 250-258.
8. Dattoli, G. & Renieri A. (1984). *Experimental and Theoretical Aspects of Free-Electron Lasers, Laser Handbook Vol. 4*, edited by M. L. Stich & M. S. Bass. Amsterdam: North Holland.
9. Dattoli, G., Renieri, A. & Torre, A. (1995). *Lectures in Free-Electron Laser Theory and Related Topics*. Singapore: World Scientific.
10. Feldhaus, J., Arthur, J. & Hastings, J. B. (2005). J. Phys. **B 38**, S799-S819.
11. *Free Electron Lasers and Other Advanced Sources of Light: Scientific Research Opportunities*, by the Committee on Free Electron Lasers and

Other Advanced Coherent Light Sources, National Research Council (National Academic Press, Washington 1994).

12. Patterson, B. D., Abela, R., Braun, H. H., Flechsig, U., Ganter, R., Kim, Y., Kirk, E., Oppelt, A., Pedrozzi, M., Reiche, S., Rivkin, L., Schmidt, T., Shmitt, B., Strocov, V. N., Tsujino S. & Wrulich, A. F. (2010). New J. Phys. **12**, 035012.

13. Huang, Z. & Kim, K. J. (2007). Phys. Rev. Special Topics Accel. Beams **10**, 034801.

14. Kim, K. J. (1986). Nucl. Instrum. Meth. **A 250**, 396-403.

15. Kim, K. J. & Xie, M. (1993). Nucl. Instrum. Methods Phys. Res. **A 331**, 359-364.

16. Brau, C. A. (1990). *Free-Electron Lasers*. Oxford: Academic Press.

17. Kondratenko, K. & Saldin, E. (1980). Part. Accel. **10**, 207-216.

18. Milton, S. V., Gluskin, E., Arnold, N. D., Benson, C., Berg, W., Biedron, S. G., Borland, M., Chae, Y. C., Dejus, R. J., Den Hartog, P. K., Deriy, B., Erdmann, M., Huang, Z., Kim, K. J., Lewellen, J. W., Li, Y., Lumpkin, A. H., Makarov, O., Moog, E. R., Nassiri, A., Sajaev, V., Soliday, R., Tieman, B. J., Trakhtenberg, E. M., Travish, G., Vasserman, I. B., Vinokurov, N. A., Wiemerslage, G. & Yang, B. X. (2001). Science **292**, 1059955.

19. Schmueser, P., Dohlus M. & Rossbach, J. (2008). *Ultraviolet and Soft-X-Ray Free-Electron Lasers*. Berlin: Springer.

20. Shintake, T. (2007). *Proc. Particle Accelerator Conference (PAC07)*. IEEE.

21. Shintake, T., Tanaka, H., Hara, T., Togawa, K., Inagaki, T., Kim, Y. J., Ishikawa, T., Kitamura, H., Baba, H., Matsumoto, H., Takeda, S., Yoshida M. & Takasu, Y. (2003). Nucl. Instrum. Methods **A507**, 382-387.

22. Altarelli, M. (2010). *From Third to Fourth-Generation Light Sources: Free-Electron Lasers in the UV and X-ray Range*, in Magnetism and Synchrotron Radiation, edited by E. Beaurepaire, H. Bulou, E. Scheurer & J. K. Kappler. Berlin: Springer.

23. Roberson, C. W. & Sprangle, P. (1989). Phys. Fluids **B1**, 3-41.

24. Hofmann, A. (2007). *The physics of synchrotron radiation*. Cambridge: Cambridge University Press.

25. Saldin, E.L., Schneidmiller, E.A. & Yurkov, M.V. (1998). Opt. Commun. **148**, 383-403.

26. Saldin, E., Schneidmiller, E. & Yurkov M. (2000). *The Physics of the Free Electron Laser*. Berlin: Springer.

27. Margaritondo, G. (1988). *Introduction to Synchrotron Radiation*. New York: Oxford.

28. Margaritondo, G. (2002). *Elements of Synchrotron Light for Biology, Chemistry, and Medical Research*. New York: Oxford.

29. G. Margaritondo and Primoz Rebernik Ribic (2011). J. Synchrotron Radiation **18**, 101-108.

30. Ayvazyan, V., Baboi, N., Bähr, J., Balandin, V., Beutner, B., Brandt, A., Bohnet, I., Bolzmann, A., Brinkmann, R., Brovko, O.I., Carneiro, J.P., Casalbuoni, S., Castellano, M., Castro, P., Catani, L., Chiadroni, E., Choroba, S., Cianchi, A., Delsim-Hashemi, H., Di Pirro, G., Dohlus, M., Düsterer, S., Edwards, H.T., Faatz, B., Fateev, A.A., Feldhaus, J., Flöttmann, K., Frisch, J., Fröhlich, L., Garvey, T., Gensch, U., Golubeva, N., Grabosch, H.-J., Grigoryan, B., Grimm, O., Hahn, U., Han, J.H., Hartrott, M.V., Honkavaara, K., Hüning, M., Ischebeck, R., Jaeschke, E., Jablonka, M., Kammering, R., Katalev, V., Keitel, B., Khodyachykh, S., Kim, Y., Kocharyan, V., Körfer, M., Kollewe, M., Kostin, D., Krämer, D., Krassilnikov, M., Kube, G., Lilje, L., Limberg, T., Lipka, D., Löhl, F., Luong, M., Magne, C., Menzel, J., Michelato, P., Miltchev, V., Minty, M., Möller, W.D., Monaco, L., Müller, W., Nagl, M., Napoly, O., Nicolosi, P., Nölle, D., Nuñez, T., Oppelt, A., Pagani, C., Paparella, R., Petersen, B., Petrosyan, B., Pflüger, J., Piot, P., Plönjes, E., Poletto, L., Proch, D., Pugachov, D., Rehlich, K., Richter, D., Riemann, S., Ross, M., Rossbach, J., Sachwitz, M., Saldin, E.L., Sandner, W., Schlarb, H., Schmidt, B., Schmitz, M., Schmüser, P., Schneider, J.R., Schneidmiller, E.A., Schreiber, H.-J., Schreiber, S., Shabunov, A.V., Sertore, D., Setzer, S., Simrock, S., Sombrowski, E., Staykov, L., Steffen, B., Stephan, F., Stulle, F., Sytchev, K.P., Thom, H., Tiedtke, K., Tischer, M., Treusch, R., Trines, D., Tsakov, I., Vardanyan, A., Wanzenberg, R., Weiland, T., Weise, H., Wendt, M., Will, I., Winter, A., Wittenburg, K., Yurkov, M.V., Zagorodnov, I., Zambolin, P., & Zapfe, K. (2006). Eur. Phys. J. D **37**, 297-303.

# THE DECISIVE ROLE OF INTERFACE PHONONS IN POLARON STATE FORMATION IN QUANTUM NANOSTRUCTURES*

A.YU. MASLOV,
O.V. PROSHINA

*Ioffe Physical-Technical Institute of the Russian Academy of Sciences,
Politechnicheskaya st., 26
Saint Petersburg, 194021, Russia*

The theory of large radius polaron is derived with consideration for the interaction of charge particles both with bulk and with interface optical phonons in the quantum nanostructures. The polaron binding energies are obtained for quantum wells, quantum wires and quantum dots. For the quantum dot case, the contributions of bulk and interface phonons have commensurable quantities. Thus the inclusion of the interface phonons reduces to some changes of polaron binding energy value. The interface phonon role is much more important in the quantum wells and wires. The essential strengthening as well as the attenuation of polaron interaction is possible with the different interrelations of the structure and barrier dielectric properties.

## 1. Introduction

In the quantum nanostructures based on materials with high ionicity the strong interaction of charge particles with polar optical phonons can lead to self-consistent bound state of an electron and phonons, that is, the large radius polaron. It has been known that the electron spectrum changes resulting from spatial quantization effect lead to an appreciable enhancement of electron-phonon interaction as the system dimensionality decreases. This is so indeed if the charge particle interaction with only one sort of optical phonons is taken into account [1]. The all-important factor is the rise of new vibration branches of optical spectrum, namely, the interface optical phonons [2] which are different for each nanostructure kind. Their role in polaron state formation is also dissimilar. The availability of the polaron states can manifest itself in multi-

---

* This work was supported by Russian Foundation for Basic Research, grant 09-02-00902-a and the program of Presidium of RAS "The Fundamental Study of Nanotechnologies and Nanomaterials" (no. 27)

phonon replicas of the electronic transition line of nanostructure optical spectra. It was observed for *CdSe* quantum dots [3,4]. The strong electron-phonon interaction was observed in cubic *ZnS* quantum wells in *ZnMgS* [5] and in undopt *CdTe/CdMnTe* quantum wells [6]. The exciton magnetic polaron in *CdMnSe/CdMgSe* quantum wells was examined in [7] using time-resolved photoluminescence.

In the present paper the theory of polaron states in the quantum wells, quantum wires and quantum dots is developed with due regard for the interaction of charge particles both with bulk and with interface optical phonons. The analytical expressions for polaron binding energies are obtained on condition that nanostructure characteristic size $r_0$ is less than the polaron radius $a_0$:

$$\frac{r_0}{a_0} < 1 \tag{1}$$

For one thing, the parameter (1) implies that the size- quantization energy of the charge particle exceeds the electron-phonon interaction energy. By this is meant that we can introduce the quantum number to classify the polaron states. This is the number of electron size quantization level obtained without regard for the electron-phonon interaction. Secondly, the parameter (1) provides the essential role of charge particle interaction with interface phonons. In "large" nanostructures where $r_0 > a_0$ this role is more less than one of the interaction with localized nanostructure phonons. The inequality (1) is satisfied for a rich variety of II-VI nanostructures where the electron-phonon interaction is strong. To determine the electron-phonon interaction intensity the polaron binding energy is considered.

## 2. Polaron states in quantum nanostructures

The total Hamiltonian of the system is given by:

$$\widehat{H} = \widehat{H}_e + \widehat{H}_{ph} + \widehat{H}_{e-ph}. \tag{2}$$

The electron Hamiltonian $\widehat{H}_e$ describes charge particle interaction with nanostructure potential, $\widehat{H}_{ph}$ contains the energies of all optical phonon branches, $\widehat{H}_{e-ph}$ is the electron-phonon interaction Hamiltonian. The parameter (1) allows us to use the adiabatic approximation to describe the polaron states. We begin with symmetric rectangular quantum well of the width $r_0$. The phonon spectrum and electron-phonon interaction parameters for such system were obtained in [8]. The symmetric interface mode frequencies are important for us are determined by solving equation:

$$\varepsilon^{(b)}(\omega) + tg\,\frac{qr_0}{2}\,\varepsilon^{(w)}(\omega) = 0. \tag{3}$$

In optical phonon frequency range, the dielectric function $\varepsilon^{(i)}(\omega)$ has the form:

$$\varepsilon^{(i)}(\omega) = \varepsilon_\infty^{(i)}\,\frac{\omega^2 - \omega_{LO}^2}{\omega^2 - \omega_{TO}^2}. \tag{4}$$

The index $i=(b),(w)$ relates to the barrier and nanostructure materials respectively. We average the total Hamiltonian of the system from Eq. (2) with fast electron motion. By unitary transformation of the averaged Hamiltonian, the energy spectrum can be found exactly. The detailed calculations are presented in [9]. The total energy of the system is written as the functional of the wave function of longitudinal electron motion in the quantum well. The minimization procedure gives the wave function and polaron binding energy. The main contribution to the polaron binding energy for the ground state of an electron can be written as:

$$E_{pol}^{(2d)} = -0.4\,\frac{me^4}{\hbar^2\left(\varepsilon_{opt}^{(b)}\right)^2}. \tag{5}$$

Here $\varepsilon_{opt}^{(b)}$ is the optical dielectric function of the barrier material, $m$ is the electron mass inside the well. One can see from Eq.(5) that the barrier dielectric properties play a crucial role in the polaron binding energy determination. It is just caused by the charge particle interaction with the symmetric interface phonon mode. The considerable polaron effect can be obtained, even though the quantum well is made from nonpolar or low-ionicity material and the barriers are made from high-ionicity material. Alternatively, the use of low-ionicity material to the barrier fabrication gives the possibility to suppress the electron-phonon interaction even though the quantum well is made of high-ionicity material. Both of these situations may be realized in the quantum wells based on II-VI/III-V hybrid structures. In the next order with parameter (1) the corrections to the Eq.(5) can be found. These corrections are related to the electron interaction both with localized phonons in the quantum well and with other branches of interface phonons [9]. The polaron radius $a_0^{(2d)}$ in a quantum well is equal to:

$$a_0^{(2d)} = \frac{\hbar^2 \varepsilon_{opt}^{(b)}}{me^2}. \tag{6}$$

Substituting material parameters [10] into Eq.(6) for the quantum well $ZnSe/CdSe$ gives us the polaron radius value $a_0^{(2d)} = 110\,\text{Å}$. The most width of the quantum well is defined by this radius. As mentioned above, this radius specifies the boundary between quantum wells with two mechanisms of polaron state formation which are qualitative dissimilar. Quasi-two-dimensional polaron states come into existence in quantum wells where the condition (1) is fulfilled. As this takes place, the polaron effect values substantially defines by he interaction of an electron with interface optical phonons. In this case the interface phonon part has a determining effect.

Likewise, the interface phonons are of great importance in the quantum wires. The polaron states can be described by the similar procedure taking into account the structure geometry. For a cylindrical quantum wire with the radius $r_0$ the interface phonon spectrum is defined by solution of the equation:

$$\frac{I'_\mu(qr_0)}{I_\mu(qr_0)}\varepsilon^{(w)}(\omega) - \frac{K'_\mu(qr_0)}{K_\mu(qr_0)}\varepsilon^{(b)}(\omega) = 0, \tag{7}$$

where $I_\mu$ is the $\mu$-th order modified Bessel function of the first kind, $K_\mu$ is the $m$-th order modified Bessel function of the second kind, $q$ is the wave vector, the quantum number $\mu$ defines different branches of the interface phonon spectrum. Our calculations show that the main contribution to the polaron binding energy gives the interface phonon interaction with $\mu = 0$.

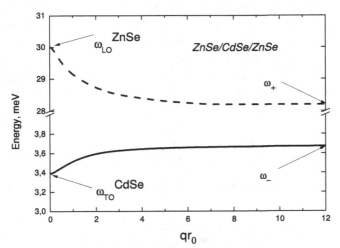

Figure 1. The wave-vector dependence of interface optical phonon energies for *ZnSe/CdSe/ZnSe* quantum wire for $\mu = 0$.

In Figure 1 is shown the wave-vector dependence of the interface phonon energies. This dependence is calculated for the quantum wire based on *CdSe* surrounded by *ZnSe* barriers with $\mu = 0$ in Eq.(7). The frequencies $\omega_+$ and $\omega_-$ are solutions of the equation $\varepsilon^{(w)}(\omega) + \varepsilon^{(b)}(\omega) = 0$. This equation is obtained from Eq.(7) over the region of large values of $q$, i.e. $qr_0 \gg 1$. The material parameters are taken from [10]. The polaron binding energy for the electron ground state has the form:

$$E_{pol}^{(1d)} = -\frac{me^4}{\hbar^2 \left(\varepsilon_{opt}^{(b)}\right)^2} \ln^2 \left(\frac{a_0^{(1d)}}{r_0}\right), \tag{8}$$

in doing so, the polaron state radius $a_0^{(1d)}$ depends on the quantum wire size obtained by the equation:

$$a_0^{(1d)} = \frac{\hbar^2 \varepsilon_{opt}^{(b)}}{me^2} \ln\left[\frac{\hbar^2 \varepsilon_{opt}^{(b)}}{me^2 r_0}\right]. \tag{9}$$

A correlation made between Eq.(6) and Eq.(9) gives that the polaron radius $a_0^{(1d)}$ for the quantum wire is just smaller than $a_0^{(2d)}$ for the quantum well. On the contrary, the polaron binding energy from is larger for the quantum wire case. For the same system *ZnSe/CdSe* the condition (1) is satisfied for the quantum wires with radius $r_0 < 40$ Å.

For the quantum dot case, with parameter (1), the electron wave function is defined by quantum dot potential. The Hamiltonian of electron-phonon interaction for such system was obtained in [11]. The electron-phonon interaction leads to the some shift of total electron state energy. It is this shift which presents electron-phonon interaction intension. Detailed analysis of this polaron shift is performed in [12]. For the spherical quantum dot the polaron energy shift is obtained as:

$$\Delta E_{pol}^{(0d)} = -\frac{e^2}{2r_0} \left(\frac{0.39}{\varepsilon_{opt}^{(w)}} + \frac{0.5}{\varepsilon_{opt}^{(b)}}\right). \tag{10}$$

We can write the polaron radius $a_0^{(0d)}$ in the form similar to Eqs.(6,9):

$$a_0^{(0d)} = \frac{\hbar^2}{me^2} \left(\frac{0.39}{\varepsilon_{opt}^{(w)}} + \frac{0.5}{\varepsilon_{opt}^{(b)}}\right)^{-1}. \tag{11}$$

It is seen from Eq.(11) that the polaron effect in the quantum dot is defined by the additive combination of polarizations of both the quantum dot and surrounding matrix materials. The contributions of bulk and interface optical phonons are the same order values. The inequality (1) is satisfied for *CdSe* quantum dot in *ZnSe* matrix when the dot radius $r_0^{(0d)} < 30$ Å.

## 3. Results and discussion

Thus the effective electron-phonon interaction may depend essentially on the phonon spectrum details for the structure being studied. The conditions have been established when the significant growth of electron-phonon interaction occurs in the quantum wells and quantum wires. An important point is that interface phonon spectrum proves to be more complicated in the quantum wire compared to the quantum well and quantum dot one.

The availability of the interface phonons leads to widening the range of materials in which the strong polaron effect should be expected. Among other things the significant electron–phonon interaction can result from the interface phonon influence in heterostructures of $Si/SiO_2$ type. The results obtained are most useful for the correct determination of optical transition energy in the semiconductor structures with strong electron-phonon interaction.

In this way an interaction of charge particles with interface optical phonons is of considerable importance in polaron state study in quantum nanostructures. It is this interaction defines the applicability conditions for the adiabatic approximation. The interface optical phonons determine the top contribution to the polaron binding energy in quantum wells and wires. The polarization properties of barrier material are a substantial part. By choosing a barrier material one can significantly affect the intensity of electron-phonon interaction and the polaron effect value.

## References

1. I.P. Ipatova, A.Yu. Maslov, O.V. Proshina, *Surf. Sci.* **507-510**, 598-602 (2002).
2. P. Halevi, in *Electromagnetic Surface Modes*, New York: John Wiley and Sons, 1982.
3. M.C. Klein, F. Hache, D. Ricard, C. Flytzanis, *Phys. Rev. B* **42**, 11123 (1990).
4. V. Yungnickel, F. Henneberger, *J. Lumin.* **70**, 238 (1996).
5. B. Urbaszek, C.M. Townsley, X. Tang, C. Morhain, A. Ballocchi, K.A. Prior, R.J. Nicholas, B.C. Cavenett, *Phys. Rev. B* **64**, 155321 (2001).
6. A.A. Dremin, D.R. Yakovlev, A.A. Sirenko, S.I. Gubarev, O.P. Shabelsky, A. Waag, M. Bayer, *Phys. Rev. B* **72**, 195337 (2005).

7. T. Godde, I.I. Reshina, S.V. Ivanov, I.A. Akimov, D.R. Yakovlev, M. Bayer, *Phys.Stat. Sol.* **247**, 1508-1510 (2010).
8. M. Mori, T. Ando, *Phys. Rev. B* **40**, 6175 (1989).
9. A.Yu. Maslov, O.V. Proshina, *Semicond.* **44**, 189 (2010).
10. Landolt-Bornstein, *Numerical Data and Functional Relationships in Science and Technology*, **17b**, Berlin Heidelberg New York: Springer-Verlag, 1982.
11. D.V. Melnikov, W.B. Fowler, Phys. Rev. B 64 (2001) 245320.
12. A.Yu. Maslov, O.V. Proshina, A.N. Rusina, *Semiconductors* **41**, 822-827 (2007).

# DIELECTRIC ANALYSIS ON OPTICAL PROPERTIES OF SILVER NANO PARTICLES IN ZRO$_2$ THIN FILM PREPARED BY SOL-GEL METHOD

Moriaki Wakaki and Eisuke Yokoyama

*Department of Optical and Imaging Science & Technology, Tokai University*
*1117 Kitakaname, Hiratsuka, Kanagawa 259-1292 JAPAN*
*e-mail: wakaki@keyaki.cc.u-tokai.ac.jp*

## 1. Introduction

The synthesis of nanosized particle is a growing research field in chemical science, in according with the extensive development of nanotechnology. The size-induced properties of nanoparticles enable the development of new applications or the addition of flexibility to existing systems in many areas [1-7]. In particular, nanoparticles of noble metals like gold and silver have been attracting more attention, because they exhibit variety of colors in the visible region based on the surface plasmon resonance. The resonance wavelength strongly depends on the size and the shape of the nanoparticles, the inter-particle distance, and the dielectric property of the surrounding medium [8-11].

In many composite material engineering researches, the aim is to design a material with desired electrical and mechanical properties. The parameters with which one can affect the macroscopic properties of the composite materials relate with the properties of the individual phases, the relative fractional volume of each phase and the shapes of the inclusions. An engineered composite material is illustrated in Fig. 1. In the system, black inclusions with permittivity $\varepsilon_i$ are randomly dispersed in a white background with permittivity $\varepsilon_e$.

The dielectric property of nanocomposites can be calculated by an effective medium approximation (EMA). Maxwell-Garnett (M-G) initiated the study of nanocomposites as he investigated the optical properties of metal colloids with minute metal spheres embedded in optically linear host material [12]. The size of the inclusions is assumed to be much smaller than the wavelength of the incident light and the composite can be treated as one homogeneous medium with effective permittivity as shown in Fig.1. In the classical M-G mixing rule, it is assumed that the local electric field on each ellipsoid is a superposition of the average external field and the average field caused by other spheres. The result for the effective permittivity is also known as the Clausius-Mossotti equation. Unfortunately, in the M-G model one is restricted to relatively small volume fraction of the inclusions because of the assumptions imposed on the model. For large volume fraction of the inclusions and for randomly intermixed constituents, Bruggeman derived EMA by considering the host material as an effective medium [13]. It assumes asymmetry between nanoparticles and matrix phases. The formula similar to the M-G equation is derived for small volume fractions of nanoparticles.

The aim of this study is the examination of the applicability of the effective medium theory to the synthesized $ZrO_2$-Ag materials. The silver nanoparticle/ $ZrO_2$ thin film composites were prepared by a sol-gel method with various silver fill fractions. The films were analyzed by a UV-Vis-NIR spectrophotometer, a transmission electron microscope (TEM) and an X-ray diffractometer (XRD). The optical absorption spectra due to the silver surface plasmon resonance were simulated using the dielectric functions reflected the M-G and the Bruggeman mixture rules.

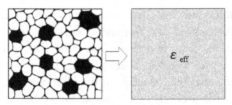

Fig. 1 Homogenization of a two phase mixture composite as the medium with effective permittivity $\varepsilon_{eff}$. Composite system is composed of black inclusions with permittivity $\varepsilon_i$ in the matrix medium with permittivity $\varepsilon_e$ .

## 2. Preparation and characterization of $ZrO_2$ thin films with silver nanoparticles

$ZrO_2$ thin films dispersed with silver nanoparticles were synthesized by the sol-gel method. The starting solution was prepared from zirconium n-propoxide, acetylacetone, 1-propanol, 2-propanol, and distilled water. The silver solution prepared from silver nitrate and diethylenetriamine. The resulting solutions were mixed with the following molar ratios; Zr : Ag = 90:10, 80:20, 70:30, 60:40, 50:50, 40:60 and 30:70. These densities of silver correspond to the volume fractions of 5.3, 11.1, 17.7, 25.0, 33.3, 42.9 and 53.9 %, respectively.

X-ray diffraction measurements were performed in a 2θ scan configuration in the range of 10-80° using an X-ray diffractometer with Cu Kα radiation (MacScience, MXP18HF). The X-ray diffraction peaks were observed at 2θ of 38.1°, 44.3°, 64.5° and 77.5° which were identified by JCPDS card as (1 1 1), (2 0 0), (2 2 0) and (3 1 1) planes of silver, respectively. A clear peak of $ZrO_2$ was not observed. It is supposed that the matrix material of $ZrO_2$ takes an amorphous structure.

TEM images of silver nanoclystallites in the zirconia films are shown in Fig. 2 for nominal Ag to Zr molar ratio [Ag]/[Zr] = 0.25, 1.00 and 2.33, respectively. The silver nanoparticles can be clearly seen, embedded in the $ZrO_2$ matrix. For the nominal Ag to Zr molar ratio [Ag]/[Zr] = 0.25 (Fig. 2a), the particles are well separated each other and their shapes are basically spherical. For Ag to Zr molar ratio [Ag]/[Zr] = 2.33 (Figure2c), the particles become coagulated and their shapes changed to an oval shape. For [Ag]/[Zr] = 1.00 (Fig. 2b), the particles show both spherical and oval shapes, and the appearance of coagulation lies in the middle in Fig. 2. With the case of any silver density, silver took a particulate state but not matrix.

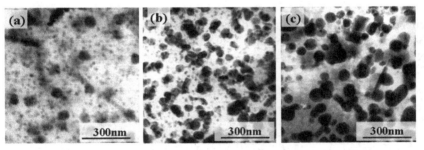

Fig.2 TEM images of $ZrO_2$-Ag films with different molar ratios of silver: (a) 80ZrO2:20Ag mol%, (b) 50ZrO2:50Ag mol% (c) 30ZrO2:70Ag mol%.

The optical absorption spectra of $ZrO_2$ thin films doped with silver nanoparticles at various densities are shown in Fig. 3. The films show an absorption band centered at about 450 nm due to the silver surface plasmon resonance. The absorption intensities become stronger as the densities of silver increase from 10 to 50 mol%, while the peak wavelength remains constant. The FWHM of the peaks remains almost constant till the Ag density of 30 mol% and becomes relatively larger above the density. On the other hand, red shift of the absorption maximum to 480 nm and the broadening of the peak were observed for the densities of silver above 60%.

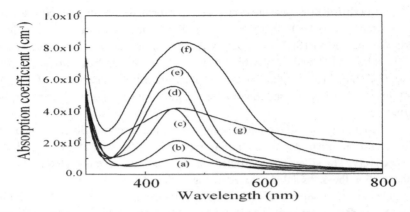

Fig. 3 Absorption spectra of ZrO₂-Ag films with different densities of silver nanoparticles: (a) 10mol%, (b) 20mol%, (c) 30mol%, (d) 40mol%, (e) 50mol%, (f) 60mol% and (g) 70mol% of silver.

## 3. Analysis on absorption spectra using effective medium approximation (EMA)

The observed absorption spectra were analyzed using EMA. In the EMA model, effective permittivity $\varepsilon_{eff}$ was described using permittivity of inclusion $\varepsilon_i$, permittivity of matrix $\varepsilon_e$, volume fraction of inclusion f and depolarization factor N. The following dielectric equation is given from the M-G model.

$$\frac{\varepsilon_e - \varepsilon_{eff}}{\varepsilon_e + K\varepsilon_{eff}} = f \frac{\varepsilon_i - \varepsilon_{eff}}{\varepsilon_i + K\varepsilon_{eff}} \quad (1)$$

$$\varepsilon_{eff} = \varepsilon_e + \frac{N\alpha}{1 - N\alpha\gamma}, \quad \gamma \equiv \frac{1}{3\varepsilon_e} + \frac{K}{4\pi\varepsilon_e}, \quad \alpha = \frac{4\pi R^3(\varepsilon_i - \varepsilon_e)}{3[\varepsilon_e + \beta(\varepsilon_i - \varepsilon_e)]} \quad (2)$$

where $R$ is the mean diameter of the nanoparticle for an ellipsoidal shape $R=(xyz)^{1/3}$, in which $x$, $y$, $z$ represent the ellipsoid semiaxis, and $\beta$, a parameter depending on the particle geometry due to depolarization factor $N$ which takes a value of 1/3 for spherical shapes, the parameter $K$ represents the rate between the electric field created at a particle position by the adjacent particles and that created by the rest of the material.

The behavior of the absorption spectrum are simulated for the ZrO₂ matrix containing spherical silver nanoparticles by the M-G model for different volume fractions f using the dielectric parameters of $\varepsilon_m$=3.57 (dielectric constant of ZrO₂ matrix), $\beta$=1/3 (spherical silver nanoparticles) and $R$=25 nm (mean diameter of Ag nanopatricles). In the case of silver densities from 10 to 40 mol%, the peak intensities become stronger keeping same spectral profile as the density of silver increases. This result shows that the dielectric property of the $ZrO_2$-Ag composite does not depend on the silver density in the low volume fraction region. On the other hand, in the case of larger silver volume fraction region such as over 60 mol%, the absorption peak wavelength shows red shift from 450 to 480 nm and the spectral profile becomes broader, which cannot be simulated by M-G model.

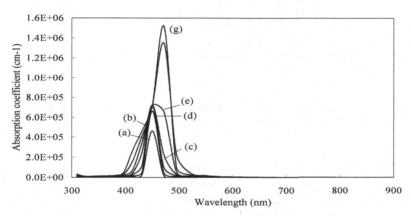

Fig. 4 Absorption spectra calculated by Maxwell-Garnett model of a ZrO2 matrix with Ag nanoparticles for different values of the volume fraction f: (a) 5.3, (b) 11.1, (c) 17.7, (d) 25.0, (e) 33.3, (f) 42.9 and (g) 53.9.

Geometrical deviations of the nanoparticles from the perfect spherical shape correspond to the deviation of β parameter from 1/3. The absorption spectra calculated for the $ZrO_2$-Ag composite with various shapes of silver nanoparticles ranging β parameter from 1/5 to 1/2 are shown in Fig. 5. These values of β parameter define the particle aspect ratio from two times to half against the incident light axis, respectively. The absorption peak wavelength strongly depends on the β value. The peak position shifts toward shorter wavelength from 580 to 380 nm as the parameter β increases.

In order to check the applicability of the M-G model for metal-dielectric composite, we tried to fit the experimental results of $ZrO_2$ thin films containing Ag nanoparticles. The particle diameter was fixed as 25nm in the calculation from TEM observation. The absorption spectrum calculated according to the M-G model is shown in Fig. 6a. The absorption peak wavelength is well fitted to the experimental result except the deference in band width. The FWHM of the experiment takes considerably larger value. This broadening of the peak is

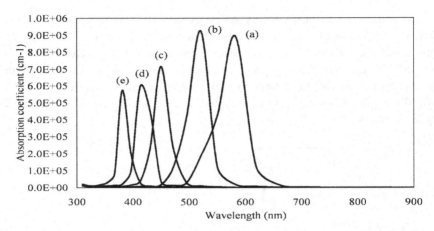

Fig. 5 Absorption spectra calculated by Maxwell-Garnett model of a ZrO2 matrix with Ag nanoparticles for different values of the shape parameter β: (a) 1/5, (b) 1/4, (c) 1/3, (d) 1/2.5 and (e) 1/2.

considered to be caused by the distribution of the particle shape and size. Much better agreement with the experimental results was obtained when the particle shape effect was considered as shown in Fig. 6a. Good fit between M-G theory and experimental spectrum was found by assuming the particle geometry distribution corresponding to β parameters from 1/5 to 1/2 as shown in Fig. 6b. The distribution of the shape was observed for the synthesized composite system using a TEM and the aspect ratio distribution of the particles from 1.0 to 2.75 was observed. This result supports the validity of the particle shape distribution used in the calculation.

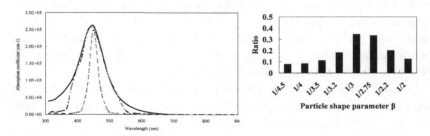

Fig. 6 (a) Experimental absorption spectum (slid line curve) fitted to the M-G model (dashed line curve) and the proposed model according to parameter value of β (dashdoted lime curve). (b) Shape distribution of silver nanoparticles using for M-G simulation.

On the other hand, the calculated spectrum using M-G model did not agree well with the experimental one in the case of larger volume fraction region over 50 mol% of silver. In general, the Bruggeman model can be applied well to the larger volume fraction region. In the Bruggeman model, the effective dielectric constant of $ZrO_2$-Ag composite is calculated by using the following equation.

$$f \frac{\varepsilon_i - \varepsilon_{eff}}{\varepsilon_i + k\varepsilon_{eff}} + (1-f)\frac{\varepsilon_e - \varepsilon_{eff}}{\varepsilon_e + k\varepsilon_{eff}} = 0 \qquad (3)$$

$$\varepsilon_{eff} = \frac{-c \pm \sqrt{c^2 + 4(1-\beta)\beta\varepsilon_e\varepsilon_i}}{4(1-\beta)}, \quad c = (\beta - f)\varepsilon_e + [\beta - (1-f)]\varepsilon_i \qquad (4)$$

The absorption spectrum calculated according to the Bruggeman model was fitted well with the experimental spectra at the volume fraction of 70 mol% of silver as shown in Fig. 7. Comparing with the M-G model, the Bruggeman model reproduce the absorption peak shifted to longer wavelength and the broadened spectral shape. However, in the case of 60 mol% of silver, both absorption the peak wavelength and the spectral shape calculated by the M-G model and the Bruggeman model did not fit well with the experimental spectrum. As a result, the applicability of M-G model is limited to smaller densities than 50 mol% and that of the Bruggeman model is over 70 mol% of silver density. It was suggested that a new model connecting the M-G model and the Bruggeman model is necessary in the middle volume fraction region of silver.

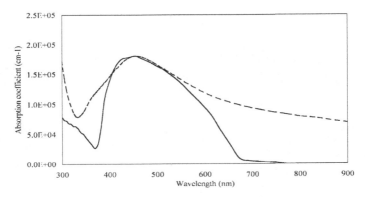

Fig. 8 Experimental absorption spectra (solid line curve) fitted to the Bruggeman model (dashed line curve).

## References

1.  S. Forster and M. Antonietti, Adv. Mater. **10** (1998) 195.
2.  M. Moffit and A. Eisenberg, Chem. Mater. **7** (1995) 1178.
3.  K. Ghosh and S. N. Maiti, J. Appl. Polym. Sci. **60** (1996) 323.
4.  R. P. Andres, J. D. Bielfeld and J. I. Henderson, Science **273** (1996) 1960.
5.  Y. Tian and T. Tatsuma, J. Am. Chem. Soc. **127** (2005) 7632.
6.  V. Subramanian, E. E. Wolf and P. V. Kamat, J. Am. Chem. Soc. **126** (2004) 4943.
7.  H. J. Jeon, S. C. Yi and S. G. Oh, Biomaterials **24**, (2003) 4921.
8.  C. A. Foss, Jr. G. L. Hornyak, J. A. Stockert and C. R. Martin, J.Phys.Chem. **98** (1994) 2963.
9.  S. K. Mandal, R. K. Roy and A. K. Pal, J. Phys. D **36** (2003) 261.
10. J. J. Mock, M. Barbic, D. R. Smith, D. A. Schultz and S. Schultz, J. Chem. Phys. **116** (2002) 6755.
11. E. Yokoyama, H. Sakata and M. Wakaki, J. Mater. Res. **24** (2009) 2541.
12. J. C. Maxwell-Garnett, Phylosophical Transactions **302** (1904) 385.
13. D. A. G. Bruggeman, Ann. Phys. **24** (1935) 636.

# TOWARDS *AB INITIO* CALCULATION OF
# THE CIRCULAR DICHROISM OF PEPTIDES

E. MOLTENI[*] and G. ONIDA

*Department of Physics, University of Milan, Milan, 20133, Italy*
*and European Theoretical Spectroscopy Facility (ETSF)*
[*]*E-mail: elena.molteni@unimi.it*

G. TIANA

*Department of Physics, University of Milan, Milan, 20133, Italy*

In this work we plan to use *ab initio* spectroscopy calculations to compute circular dichroism (CD) spectra of peptides. CD provides information on protein secondary structure content; peptides, instead, remain difficult to address, due to their tendency to adopt multiple conformations in equilibrium. Therefore peptides are an interesting test-case for *ab initio* calculation of CD spectra. As a first application, we focus on the (83-92) fragment of HIV-1 protease, which is known to be involved in the folding and dimerization of this protein. As a preliminary step, we performed classical molecular dynamics (MD) simulations, in order to obtain a set of representative conformers of the peptide. Then, on some of the obtained conformations, we calculated absorption spectra at the independent particle, RPA and TDLDA levels, showing the presence of charge transfer excitations, and their influence on spectral features.

## 1. Introduction

This work is aimed towards further developments of our presently available *ab initio* theoretical spectroscopy tools suitable for studying and understanding correlations between optical spectra and conformational changes in molecules of biological interest.

The importance of structure for protein function is well known, and also the processes through which a protein assumes its native conformation (folding), or a "wrong" one (misfolding) have long been a subject of study, from both an experimental and a computational point of view.

Secondary structure (ss) elements, such as α-helix, β-sheet etc., can be considered as the basic building blocks of tertiary structure, *i.e.* of the protein 3D conformation. They are determined by the arrangement of backbone amide groups, and often characterized by specific patterns of hydrogen bonds and backbone dihedral angles.

Experimentally, ss elements can be identified by optical techniques, in particular circular dichroism (CD), where the various secondary structure types give distinct "fingerprint" spectral features. Methods and softwares relying on databases of proteins of known structure can then be used to analyze CD spectra.[1] These methods are strongly phenomenological, therefore not transferrable to different systems. In particular, they cannot be applied to peptides, which are highly flexible and tend to assume an ensemble of conformations in solution, rather than a single structure. On the other hand, studying peptide conformation can provide valuable information, particularly for peptides which are known to play a key role (either structural or functional) in the protein to which they belong.

A computational approach to this issue may help in interpreting experimental data, with the advantage of not relying on *ad hoc* criteria for defining secondary structure (as it happens instead in classical molecular mechanics force fields).

## 2. Circular dichroism

CD is due to the difference in molar extinction between left and right circularly polarized light by a chiral medium, and the experimentally measured macroscopic effects can be related to this tensor:[2,3]

$$\mathcal{G}_{\mu\nu}(\omega) = \frac{1}{\hbar} \sum_{n \neq 0} \left\{ \frac{< 0|\hat{\mu}_\mu|n><n|\hat{m}_\nu|0 >}{\omega_{no} - \omega - i\gamma_{no}} + \frac{< 0|\hat{m}_\nu|n><n|\hat{\mu}_\mu|0 >}{\omega_{no} + \omega + \gamma_{no}} \right\}$$

(1)

where

$$\hat{\mu} = q\hat{r}$$

(2)

and

$$\hat{m} = \frac{q}{2m}\hat{r} \times \hat{p}$$

(3)

are the electric and magnetic dipole moments, respectively.

The calculation of CD spectra has been recently implemented in some *ab initio* plane wave codes,[2] but it is not yet of common use. Moreover, in plane wave codes, one must be careful regarding the issue of gauge invariance. On the other hand, to be able to compute CD (due to difference in molar extinction), one first has to compute molar extinction itself, that is to say, absorption spectra.

This can be seen also from Eq. (1), where the matrix element of the electric dipole moment is one of the "ingredients". Therefore we start by computing optical absorption spectra, before moving to the more complex CD spectra.

## 3. Methods and investigated systems

As a first application, we focus on the 83-92 fragment of HIV-1 protease. This is a 10 aminoacid peptide, closely related to the more studied 83-93 fragment, which is known to be involved in the folding and dimerization of HIV-1 protease, and to be a possible folding inhibitor of this protein.[4,5] The 83-92 peptide assumes a partially α-helical conformation when it is in the protein: this makes it an interesting test case, since α-helix is one of the best identifiable secondary structure elements from the point of view of circular dichroism.

In order to obtain possible conformations of the peptide in solution, we performed a classical molecular dynamics (MD) simulation in explicit water with the replica exchange (RE) technique, using the GROMACS code.[6] The RE method enhances conformational sampling by allowing temperature exchange between different copies of the system (called replicas), simulated at different temperatures. High T replicas have high kinetic energy and they are therefore able to sample larger areas in phase space. Exchanging configurations with replicas at lower T, this method avoids letting the "cold" systems be trapped in local energy minima, without the artifacts which would arise if performing a single simulation at very high T.[7]

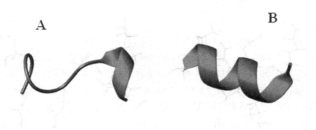

A      B

Figure 1. Conformations 1 (A) and 2 (B) chosen for *ab initio* calculations.

The obtained conformations were grouped in clusters of similar structures, based on backbone RMSD (root mean square deviation of atomic positions). For *ab initio* calculations we chose two different conformations, the first belonging to the most populated cluster, the second completely α-helical (Fig. 1), to test sensitivity to conformational changes. On these two conformations, we calculated absorption spectra at the independent particle (IP), RPA and TDLDA levels, using the ABINIT and YAMBO codes.[8,9]

## 4. Results

The first features one can observe in spectra calculated at the independent particle level (Fig. 2) are their anisotropy and dependence on conformation.

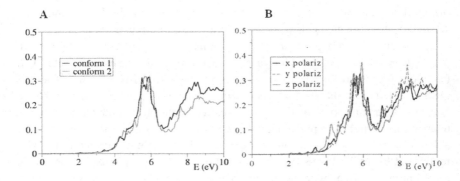

Figure 2. Independent particle absorption spectra of two chosen conformations of the 83-92 fragment of HIV-1 protease, each averaged on the three possible light polarization directions (A), and spectra of conformation 1, for the three polarization directions (B).

By comparing these spectra with the density of states (DOS) (Fig. 3A,B), we see that the spectrum onset is around 2 eV, while the energy gap is much smaller (1 eV for conformation 1, and 0.64 eV for conformation 2). This apparent anomaly can be explained by direct inspection of the wavefunctions near the highest occupied molecular level (HOMO). In this energy region, most occupied (Fig. 3C) and empty (Fig. 3D) electronic states are localized far from each other, in two different parts of the molecule, and this happens for both conformations. Therefore many near-gap transitions have a very small dipole matrix element, because of small spatial superposition.

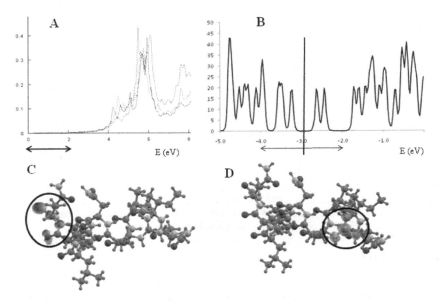

Figure 3. Conformation 2: a part of the independent particle absorption spectrum (the three possible light polarizations), highlighting spectrum onset (A); density of states, highlighting Fermi energy and a range of 2 eV around it (B); wavefunction localization for the HOMO (C) and HOMO + 3 (D).

This phenomenon of charge transfer excitations is known to be not very well described by TDLDA;[10] in fact also in our case TDDFT spectra disagree with experimental data (Fig. 4A). The sharp onset of the experimental spectrum suggests excitonic nature of this peak, which, together with the charge transfer character of the transitions around the Fermi level, explains the bad performance of TDLDA. Despite this problems, the observed anisotropy and conformation dependence are still present also at the RPA level (Fig. 4B).

It remains an open question, whether the best strategy to overcome this limitation will involve going to a higher level, *i.e.* using a nonlocal kernel or switching to a many-body perturbation theory approach as GW+BSE, or if, on the contrary, this is a case where stopping at the independent quasiparticle level, maybe applying a rigid blueshift, could be a reasonable alternative for computing absorption spectra and, in perspective, also CD spectra.

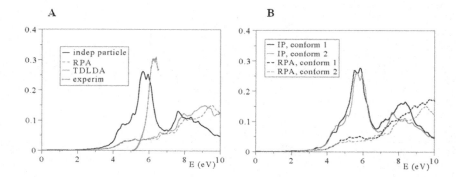

Figure 4. Independent particle (IP), RPA and TDDFT spectra of conformation 2, superimposed to a representative experimental spectrum (A); IP and RPA spectra for conformation 1 and 2. In both cases calculated spectra are averaged on the three polarization directions.

## 5. Conclusions

In the present work we calculated optical absorption spectra of the 83-92 fragment of HIV-1 protease with DFT methods, after obtaining a set of possible conformations of the peptide by classical molecular dynamics simulations. This is a preliminary step to the implementation and use of circular dichroism equations in the ABINIT code,[8] focusing in particular on molecules of biological interest. Already from calculations at the independent particle level, we observed anisotropy of absorption spectra and their dependence on peptide conformation. Moreover, by analyzing wavefunction localization in the HOMO-LUMO energy region, we found that the peptide under study exhibits charge transfer excitations, which strongly affect its calculated spectra, explaining the position of the absorption onset, quite higher in energy than one would expect on the basis of the joint DOS. This latter feature shows qualitative agreement with experimental absorption spectra of the same peptide.

## Acknowledgments

We wish to thank CINECA for computing time, and D. Varsano for useful discussions.

## References

1. L. Whitmore and B. A. Wallace, *Biopolymers* **89**, 392 (2008).
2. D. Varsano, L. A. Espinosa-Leal, X. Andrade, M. A. L. Marques, R. di Felice and A. Rubio, *Phys. Chem. Chem. Phys.* **11**, 4481 (2009).

3. F. Hidalgo, A. Sánchez-Castillo and C. Noguez, *Phys. Rev. B* **79**, 075438 (2009).

4. R. A. Broglia, D. Provasi, F. Vasile, G. Ottolina, R. Longhi and G. Tiana, *Proteins* **62**, 928 (2006).

5. G. Verkhivker, G. Tiana, C. Camilloni, D. Provasi and R. A. Broglia, *Biophys. J.* **95**, 550 (2008).

6. B. Hess, C. Kutzner, D. van der Spoel and E. Lindahl, *J. Chem. Theory Comput.* **4**, 435 (2008).

7. Y. Sugita and Y. Okamoto, *Chem. Phys. Lett.* **314**, 141 (1999).

8. X. Gonze et al., *Computer Phys. Commun.* **180**, 2582 (2009).

9. A. Marini, C. Hogan, M. Grüning and D. Varsano, *Comp. Phys. Comm.* **180**, 1392 (2009).

10. A. Dreuw, J. L. Weisman and M. Head-Gordon, *J. Chem. Phys.* **119**, 2943 (2003).

# DYNAMICS AND SPECTRAL PROPERTIES OF FREE-STANDING NEGATIVELY-CURVED CARBON SURFACES

## M. De Corato[1] and G. Benedek[1,2]

[1]*Dipartimento di Scienza dei Materiali, Università di Milano-Bicocca,*
*Via R. Cozzi 53, 20125 Milano, Italy*
[2]*Donostia International Physics Center (DIPC), EHU-UPV*
*P. Manuel de Lardizàbal 4, 20018 Donostia/San Sebastian, Spain*

The growing interest for the physical properties of graphene and its chemical functionalization is triggering a renewed attention to other more complex forms of $sp^2$ nanostructured carbon, e.g., carbon schwarzites. The structural complexity of these forms prevents in many cases the possibility of ab-initio calculations of their electronic and vibrational properties. On the other hand in systems with a certain level of conjugation many of these properties depend primarily on the topological features of the structure – notably on the graph describing the bonding network and the associated adjacency matrix. In this work, specifically devoted to planar schwarzites, it is shown that their vibrational spectrum as obtained from the diagonalization of the adjacency matrix (*topological phonons*) reflects the main features of *ab-initio* calculations. Thus the method is suitable to the study of thermodynamic and spectral phonon properties of defective schwarzites which involve integrals over the unperturbed phonon density of states. Calculations of the phonon density of states smallest planar schwarzites and the effects of various type of local perturbation are presented and thoroughly discussed.

## 1. Introduction

There are atomic surfaces which have no underlying bulk and are free-standing thanks to their covalent bonding architecture. Their vibrational structure reflects, in its general features, their topological constitution, thus playing a relevant role in the growth mechanisms and spectroscopic characterization. The recent Nobel Prize awarded to the studies on graphene [1-5] has, by extension, revived the interest in the vast zoo of curved surfaces of carbon which are made possible by sp2 hybridization. Besides the well-known forms like fullerenes [6], single-walled and multi-walled nanotubes [7], worth mentioning are the three-dimensional forms of 2p2 carbon, random schwarzites. Figure 1(a,c) shows a transmission electron microscope (TEM) image of a random carbon schwarzite obtained by supersonic cluster beam deposition with a deposition energy of 0.1 eV/atom [8-10]. Raman and near-edge x-ray absorption fine structure (NEXAFS) spectra indicate a pure sp2 bonding structure, suggesting a single, highly connected graphene sheet with an average pore diameter in the range of 100 nm. Although carbon schwarzites have been first synthesized and characterized a decade ago [8-10], they did not know the glamour of the ordered sp2 carbon forms. Nevertheless random schwarzites, otherwise termed spongy carbon [9], qualify for unique properties, such as, e.g., unconventional magnetism [11,12], and applications in efficient supercapacitors [13], field emitters [14-16], carbon-based composites [17], up to the recent the recent demonstration of interfacing live cells with nanocarbon substrates [18].

Triply periodic minimal surfaces as possible sp2 carbon structures have been theoretically suggested already in the mid-eighties [19], then with more momentum in the early nineties, following the nanotube vogue [20-25]. They have since termed schwarzites after the name of the mathematician Hermann Schwarz [26] who first investigated that class of surfaces. The synthesis of random schwarzites [9] was obtained by means of supersonic cluster beam deposition (SCBD) [27]. SCBD experiments demonstrated that spongy carbon grows in presence of finely dispersed Mo nano-catalysts, with a porous size decreasing with increasing deposition energy and no tendency to form triply periodic structures. These aspects, as well as the growth in the form of a self-affine minimal random surface, have been theoretical elucidated on the basis of pure topological arguments [10, 28, 29]. Many relevant properties of schwarzites can actually be derived in a first approximation from a topological analysis. For a thorough discussion of these aspects the reader is referred to recent book chapter [30].

In this paper it is shown that also the vibrational spectra of schwarzitic structures can be estimated from topology, more precisely from the adjacency matrices. After assessing the method on standard cases as the fullerene C60 and the simplest three-periodic schwarzite fcc-(C28)2, for which the vibrational spectra are well established, a novel class of two-periodic schwarzites shall be introduced and their approximate vibrational spectra will be derived with the adjacency matrix method. The interest for two-periodic schwarzites, here discussed for the first time, is related to the possibility of growing them by means of near-to-come planar technologies through the direct joining of nanotubes.

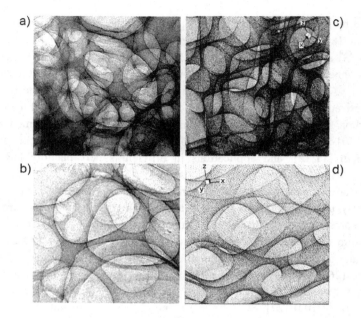

Fig. 1: Two transmission electron microscope (TEM) pictures (a,b) of a random carbon schwarzite as grown by supersonic cluster beam deposition [10] and two simulations (c,d, respectively) of the TEM images obtained from analytical approximations of three-periodic minimal surfaces (D-type (c) and P-type (d) schwarzites) with a self-affine distortion [30].

## 2. Adjacency matrix

The topological description and classification of P- and D-type schwarzites, whether three-periodic or random, have been summarized in Ref. [30] and will not be further discussed here. In this paper we shall consider both three-periodic and two-periodic schwarzites of the archimedean type, i.d., originating from tiling a three- or two-periodic surface with only two types of polygons, heptagons and hexagons.

By virtue of the Euler theorem on tiling, an element of a three-periodic D-type schwarzite, having each four joints with equivalent elements, must have 12 heptagons and any given number $h$ ($\neq 1$) of hexagons. The corresponding unit cell, made of two elements, contains therefore

$$N = 2 \cdot (12 \cdot 7 + 6h)/3 = 2 \cdot (28 + 2h) \text{ atoms} \qquad \text{(3D)} \qquad (1)$$

and corresponds to the formula $(C_{28+2h})_2$.

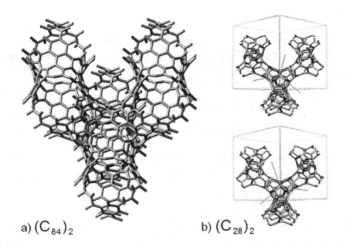

a) $(C_{84})_2$    b) $(C_{28})_2$

Fig. 2: Four unit cells of the D-type carbon schwarzites (C84)2 (a) and (C28)2 (b). The former is the smallest D-type schwarzite where no abutting 7-membered rings occur [25,30], and is therefore analogous to the buckyball C60 which is the smallest fullerene with no abutting 5-membered rings. The latter is the smallest D-type schwarzite with no 6-membered ring but only 7-membered rings (12 per element, i.e., 24 per unit cell), analogous to the fullerene C20, which is only made of twelve 5-membered rings. It exists in two chiral enantiomers, as signalled by the orientation of the red lines in (b) [30].

Similarly an element of a two-periodic schwarzite, having each three joints with equivalent elements all lying on a plane, must have 6 heptagons and any given number $h$ ($\neq 1$) of hexagons. Thus the corresponding unit cell, made of two elements and having therefore four joints with four adjacent unit cells, contains

$$N = 2 \cdot (6 \cdot 7 + 6h)/3 = 2 \cdot (14 + 2h) \text{ atoms} \qquad (2D) \qquad (2)$$

and corresponds to the formula $(C_{14+2h})_2$.

The structural and physical properties which exclusively depend upon the topological constitution of the carbon network can be evidenced, and separated from those properties which are determined by the actual local bonding, by viewing the three-coordinated sp$^2$-bonded carbon networks as graphs. Then the topology of graphs, notably their topological indices like the local and global Wiener indix, the compactness index, etc., can be used in order to rank the isomers with respect to properties like stability, site reactivity, etc. The first graph-theory ingredient for this kind of analysis is the adjacency matrix (AM), which is discussed in the next Section

118

## 2. The adjacency matrix

From the topological point of view a pure carbon network is completely assigned by the set of bonds linking pairs of neighbor atoms labelled by $i$ and $j$. The adjacency matrix $A_{ij}$, defined as

$$A_{ij} = \begin{cases} 1 & \text{if atoms } i, j \text{ are connected by a bond} \\ 0 & \text{otherwise, including } i = j \end{cases} \tag{3}$$

provides full information on the network connectivity and allows, through its eigenvalues and eigenvectors, to obtain an approximate description of certain physical properties of the network.

The eigenvectors of the adjacency matrix, for example, lead to the definition of topological coordinates of three-coordinated carbon structures like fullerenes [31], nanotubes [32], and schwarzites [33, 34]. The topological coordinates essentially consist of sets of atomic positions connected according to the adjacency matrix which have the highest point symmetry compatible with the adjacencies. In practice the topological coordinates of a given isomer can provide a good starting set of cartesian-coordinate positions for stuctural optimization based either on classical or quantum molecular dynamics.

Another application of the adjacency matrix is the calculation of the electronic energies of a mono-atomic network in the tight-binding (TB) approximation for a band originated from a single atomic state, e.g., the $p_z$ band in a sp$^2$ carbon network. By assuming the same diagonal matrix element $\alpha$ of the Hamiltonian for all atomic orbitals, the same overlap integral $s$ and the same Hamiltonian matrix element (resonance integral) $\beta$ between the atomic orbitals for all nearest neighbor pairs, the energy eigenvalues $E = E_j$ and the eigenvectors $\mathbf{c} = \mathbf{c}_j$ providing the coefficients of atomic orbital combinations are obtained by solving the TB equation

$$(\mathrm{I} - s\,\mathrm{A})^{-1}(\beta\,\mathrm{A} + \alpha\,\mathrm{I})\mathbf{c} = E\,\mathrm{I}\mathbf{c} \tag{4}$$

where I is the unitary matrix. The extension of this equation to the periodic schwarzite lattice would provide the valence band structure. In this way a qualitative information about the size of the orbital gap between the highest occupied (HOMO) and lowest unoccupied (LUMO) molecular orbitals can be obtained as a function of the topology, here represented by the adjacency matrix, and to infer whether the periodic schwarzite will be an insulator or a metal. As shown in Ref. [30], the smallest D-type schwarzites of tetrahedral symmetry have alternatively metal or insulating character for an increasing number $h$ of hexagons in the unit cell, the limit for $h \to \infty$ being a graphenic semi-metal.

For three-coordinated $sp^2$ a similar approach permits to obtain an approximate description of the phonon spectra which is useful, e.g., in the calculation of phonon-dependent integrated properties like the vibrational entropy and other thermodynamic quantities. An example is the calculation of the average porosity of SCBD-grown random schwarzites as a function of the deposition energy [30]. In this case it is shown that the vibrational contribution does not allow to reduce the average pore size below a certain minimum. In other words the theory indicates an optimal deposition energy per atom at which the maximum in the specific surface area of the carbon foam is obtained. In the next Section the discussion shall be focussed on the vibrational spectra relevant to the infrared spectroscopy of complex carbon surfaces as directly derived from the adjacency matrix.

## 3. Topological phonon spectra of curved $sp^2$ carbon surfaces

The vibrational spectra extracted from the adjacency matrix of a D-type schwarzite unit cell with each of the six terminations closed on the opposite one (so as to form a three-handle torus) are topologically equivalent to the spectra at zero wavevector ($q = 0$) of the corresponding three-periodic solid with periodic boundary conditions. For a flat surface (graphene) the vectorial nature of the phonon displacement field at $q = 0$ is factorized into a transverse out-of-plane component, corresponding to a transverse optical mode normal to the surface ($TO_\perp$ mode) having a frequency $\omega_\perp = 16.3$ rad/s, and two orthogonal in-plane components corresponding to the parallel optical ($TO_\parallel$) and longitudinal optical (LO) modes, having the same frequency of $\omega_\parallel = 29.8$ rad/s. The degeneracy of $TO_\parallel$ and LO modes at $q = 0$ is intrinsically due to the symmetry of the three $sp^2$ bonds of forming three angles in plane of $120°$.

On curved surfaces the three angles are in general distorted and no longer in plane – a fact irrelevant, however, at the topological level. The assumption that each of the orthogonal components at each atom only couples with the same component of the three adjacent atoms reduces the dynamical problem to the diagonalization of three independent combinations of the adjacent matrix. Actually only two nearest neighbor force constants $f_\perp$ and $f_\parallel$ are needed. We use for this approximation the terms *topological dynamics* and *topological phonons*. The eigenvalue equation providing the angular frequencies $\omega = \omega_{\alpha v}$ and the components $u_i = u_{i\alpha,v}$ of the atomic displacements for each phonon $v$ and each polarization $\alpha = \perp, \parallel$ can be expressed in terms of the adjacency matrix as

$$-M\omega^2 u_i = f_\alpha \sum_j (A_{ij} - 3\delta_{ij})u_j, \qquad \alpha = \perp, \| \qquad (5)$$

The force constants $f_\alpha$ can be fitted to the respective frequencies $\omega_\perp$ and $\omega_\|$ for graphene and considered to be transferable to other $sp^2$ carbon structures. In Eq. (5) $M$ is the carbon atom mass, and the term with the Kronecker delta is implied by the translational invariance of the system hamiltonian.

The angular frequencies are directly obtained from the eigenvalues $\lambda_{\alpha v}$ of the adjacency matrix as

$$\omega_{\alpha v} = \left[ \frac{f_\alpha}{M}(3 - \lambda_{\alpha v}) \right]^{1/2}. \qquad (6)$$

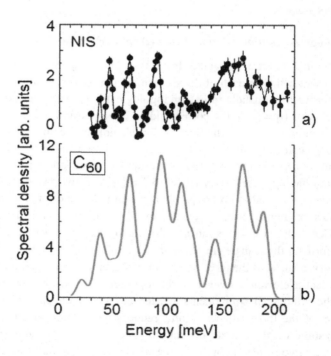

Fig. 3: Comparison between the vibrational spectrum of the icosahedral fullerene C60 measured with neutron inelastic scattering (NIS) [35] (a) and the topological phonon spectrum calculated from the adjacency matrix (b). In the latter spectrum a finite linewidth is attributed to each phonon peak in order to obtain a smooth spectrum with a resolution comparable to that of experiment.

An example of topological phonon spectrum calculated with Eq. (5) is shown in Fig. 3 for the icosahedral fullerene $C_{60}$ and compared with the experimental spectrum obtained with inelastic neutron scattering (NIS) [35, 36]. The good correspondence of the NIS spectrum peaks to those of the topological phonon spectrum indicates that the gross features of the $C_{60}$ vibrational spectrum are accounted for by its topology, i.e., by its bonding network. Similar results have been obtained for the D-type schwarzite $(C_{28})_2$ for which a comparison is possible between the *ab-initio* and the topological phonon spectra [34]. On this basis calculations of the topological phonon spectra for the smallest planar schwarzites at zero wavevector, i.e., relevant to vibrational spectroscopy, are presented in the next section, together with an analysis of the effects of various types of defects which may occur in functionalized planar schwarzites.

## 4.   Planar G-type schwarzites

The simplest of planar schwarzites is given by Eq. 2 with $h = 0$ and has the unit-cell formula $(C_{14})_2$. It is a tiling with six 7-membered rings per element of a two-periodic surface having the geometry of a graphene network (Fig. 4), with the unit cell made of two elements, each element being linked to three identical elements through 8-membered rings. Larger planar schwarzites organized in a graphene network can be generated by the insertion in each element of a number $h$ of 6-membered rings disposed in a suitable way so as to preserve the trigonal symmetry. Actually each element of Fig. 4 may act as a joint of three nanotubes of equal length of the type (4,0). By analogy with the three-periodic D-type schwarzites, which are organized as a diamond lattice, we shall call the present class of two-periodic schwarzites organized as a graphene sheet G-type schwarzites.

The unit cell of the smallest G-type planar schwarzite $(C_{14})_2$ (Fig. 4) is obtained by -joining two elements $C_{14}$ (A and B). Since the joint requires either a 45° or a -45° rotation of B with respect to A around the axis joining the two cluster centers, the unit cell acquires a chirality and so is for the planar schwarzite, once all elements are joined with the same rule. Thus $(C_{14})_2$ has two chiral enantiomers, like the smallest three-periodic D-type schwarzite $(C_{18})_2$ . These elements $C_{14}$ can however be joined also through (4,0)-nanotubes of various lenghts (Fig. 5): for example, in $(C_{18})_2$ one single (4)-cyclacene is inserted between the two elements of the original unit cell, which breaks the trigonal symmetry and leads to a buckled structure. On the contrary in $(C_{26})_2$ the insertion of one (4)-cyclacene into each neck removes the chirality and yields a graphene-like hexagonal network, i.e. to a G-type schwarzite.

122

$(C_{14})_2$

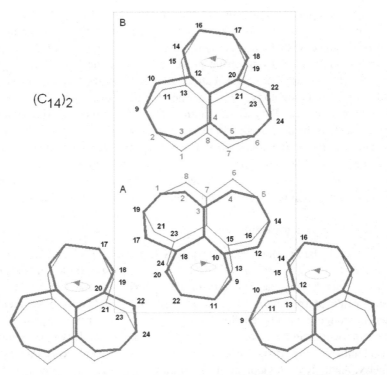

Fig. 4: The smallest G-type planar schwarzite (C14)2 exclusively made of 7-membered carbon rings. The unit cell is obtained by joining according to the corresponding numbers two elements C14 (A and B) through a neck 8-membered ring. Since the joint implies either a 45o or a -45o rotation of B with respect to A around the axis joining the two cluster centers, the unit cell acquires a chirality and so is for the planar schwarzite, once all elements are joined with the same rule. Thus (C14)2 has two chiral enantiomers, like the smallest three-periodic D-type schwarzite (C18)2.

There are quite a few theoretical examples reported in the literature of large planar G-type schwarzites aimed at describing the local structure of experimentally observed Y-joints between single-wall carbon nanotubes as well as more complex nanotube structures. We just mention some early predictions by Chernozatonskii [37] and Spadoni *et al* [38], as well as a recent more extensive paper by Romo-Herrera *et al* [39] and references therein. It has been noted by Spadoni *et al* [38] that, given any schwarzite with $N$ atoms per unit cell, elements of a G-type schwarzite with $14 + 2h$ atoms (Eq. (2)), which coordinate three identical elements, can be joined each other in the same way atoms are in the original schwarzite. Such a replacement algorithm yields a new schwarzite having $N(14+2h)$ atoms per unit cell. A further iteration of the replacement algorithm gives another schwarzite with $N(14+2h)^2$ atoms per unit cell, etc. If $a_0$ and $a_1$ are the interatomic distance and the distance between two neighbor elements, respectively, the volume of the unit cell increases after $n$ terations by a factor $(a_1/a_0)^{3n}$ whereas the number of atoms increases by the factor $(14 + 2h)^n$. The iteration *ad infinitum* yields a schwarzite which has no periodicity but acquires the character of a *topological fractal* [40,41] with a dimension

$$D = \ln(14 + 2h) / \ln(a_1 / a_0) . \tag{7}$$

For $h = 0$ (Fig. 4) the ratio $a_1/a_0$ must be larger than 2.5, so that the replacement algorithm in either a three-periodic or a planar two-periodic schwarzite would yield a fractal carbon foam with $D < 1.723$. For $h > 1$ the fractal dimension does not necessarily increase because also the distance $a_1$ between two neighbor elements increases. For example, in the case where 6-membered rings are

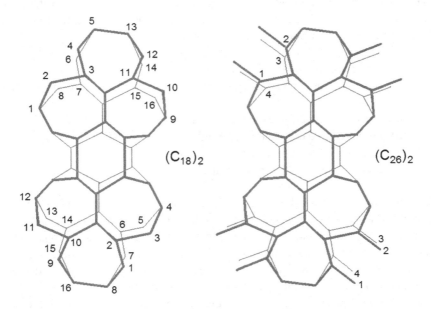

Fig. 5: The elements C14 of the smallest G-type planar schwarzites can be joined either directly as in (C14)2 or through (4,0)-nanotubes of various lenghts. E.g., in (C18)2 one single (4)-cyclacene is inserted between the two elements of the original unit cell, which breaks the trigonal symmetry and leads to a buckled planar structure. On the contrary in (C26)2 the insertion of one (4)-cyclacene into each neck removes the chirality and yields a graphene-like hexagonal network.

inserted to make longer necks connecting the trigonal joints, as, e.g., in $(C_{26})_2$ (Fig. 5) where a (4)-cyclacene is added at each neck, $h$ increases by 6 units whereas $a_1/a_0$ increases by 1 for each (4)-cyclacene added at each neck. Thus for long (4,0) nanotubes made of $m$ cyclacene rings connecting the trigonal $C_{14}$ joints the fractal dimension decreases, as expected, logarithmically towards unity like $1 + \ln(12)/\ln(m)$.

## 5.  Vibrational spectra of pure and defective planar schwarzites

The topological phonon spectra at vanishing wavevector for the smallest planar schwarzites described in Figs. 4 and 5, and the perturbing effects of a variety of defects are now discussed on the basis of the diagonalization of the adjacency matrix, Eqs. (5,6). Since we are interested in the spectra at $\mathbf{q} = 0$ the calculations are made for a single unit cell with periodic boundary conditions, which from the topological point of view correspond to connections of opposite terminations so as to generate a two-handle torus. In this respect the three planar schwarzites here considered are topologically equivalent since they are closed on the same number of handles and have therefore the same number of 7-membered rings. The following structures and defects have been considered:

(a)  the perfect $(C_{14})_2$, $(C_{18})_2$ and $(C_{26})_2$ lattices;
(b)  break of one bond connecting one of two central atoms on the trigonal axis;
(c)  one of the three force constants of a central atom is doubled;
(d)  all three force constant of a central atoms are doubled;
(e)  the mass of one central atom has been multiplied by 4;
(f)  a vacancy at the site of a central atom.

The results are collected in Figs. 6-8. The effects of these kinds of force-constant and mass perturbations, as appear from the comparisons of spectra (b-f) with the spectrum of the corresponding perfect structures are quite transparent. They are entirely similar to those occurring in three-dimensional insulators and semiconductors, which have been extensively investigated with detailed dynamical models in the late sixties and early seventies [42].

The two-dimensional nature of schwarzite surfaces has a limited relevance, the topological effects being dominant. In this respect the calculation of the defect perturbation of the phonon density of states (DOS) based on adjacency matrix approximation is reliable, as long as the general features are concerned. A formal reason is that the perturbing effects of the spectrum depend mostly on the Hilbert transforms of the bulk densities projected onto the defect sites, i.e., on the real parts of the projected Green's functions: they are integral functions and essentially depend on the gross features of the unperturbed spectra. In detail it appears that the stiffening of the force constants leads to localized modes

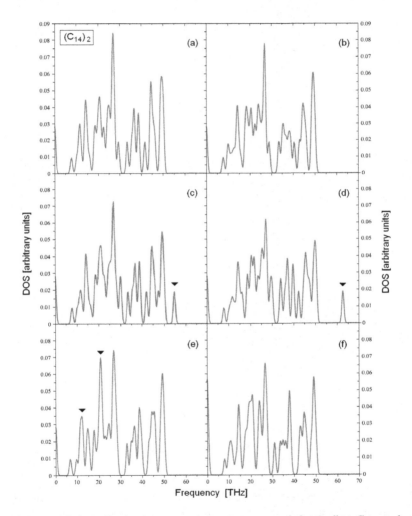

Fig. 6: The topological vibrational spectra at zero wavevector of the smallest G-type planar schwarzite (C14)2: (a) pure lattice; (b) one bond broken at a central atom; (c) one of the three force constants connecting a central atom is doubled; (d) all three force constants of a central atom are doubled; (e) the mass of a central atom is multiplied by 4; (f) a vacancy at a central atom site.

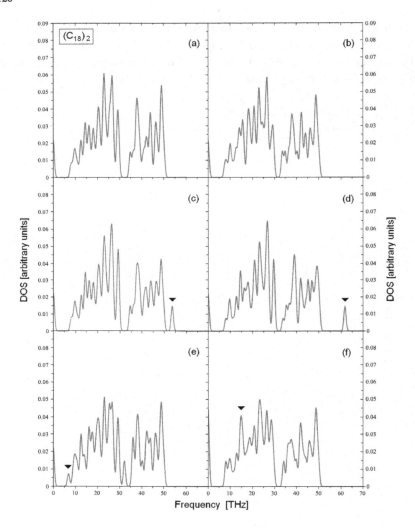

Fig. 7: Same as in Fig. 6 for (C18)2.

above the maximum frequency of the unperturbed spectrum (arrows in Figs. 6-8(c,d)), whereas the break of bonds and the mass increase produce a general phonon softening, with the emergence of resonances in the lower part of the spectrum (arrows in Figs. 6-7(e)). In all structures there is a narrow gap around 32 THz separating the lower phonon band of shear vertical (ZO) modes from the upper longitudinal (LO) and shear horizontal (SH) phonon bands. In some case the local perturbation causes the appearance of a gap mode, as for example for a vacancy in $(C_{14})_2$ (Fig. 6(f)) or a mass increase perturbation in $(C_{18})_2$ (Fig. 7(e)).

These few examples should illustrate the optical vibrational spectra which are expected for this class of graphene-like nanostrustrured carbon, and the spectral modifications induced by defects, or in general by any local structure which may occur, for example, in functionalized $sp^2$ carbon samples.

## 6. Conclusions

The theoretical study of the vibrational spectra of complex $sp^2$-bonded carbon surfaces with promising applications in various areas of nanotechnology is in most cases inaccessible to present first-principle methods. Moreover the presence of conjugation emphasizes the overall (*holistic*) dynamical features of the system over the local ones, and the former are largely determined by topology. It has been recognized for this class of materials that the vibrational spectra directly derived from the adjacency matrix reproduce fairly well the general spectral features known from either experiment (Fig.3) or ab-initio calculations [34]. Thus the vibrational properties of $sp^2$-carbon surfaces which depends on integrals over the spectrum, such as their thermodynamic functions [30] or defect-induced perturbations of the phonon density, can be safely evaluated on pure topological grounds, once the bonding network is known.

A few examples have been given for the smallest planar carbon schwarzites. This particular class of materials, though purely hypothetical at present, has been chosen because the elementary constituents of trigonal symmetry may work as Y-joints for the practical realization of planar nanotube networks, and a spectroscopic characterization of the structure is envisaged. As a follow-up of the original work on three-periodic schwarzites by Vanderbilt and Tersoff [25], who have presented these labirinth structures as plumber's nightmares, a nano-plumbing where nanotubes can be joined so as to form schwarzitic networks for various purposes is conceivable today, especially is two dimensions with the methods of planar technology on a suitable substrate.

## 7. Acknowledgements

One of us (G.B.) thanks Prof. Wanda Andreoni and Dr. Fabio Pietrucci for many useful discussions during a recent visit at CECAM, Ecole Polytechnique Fédérale de Lausanne (EPFL), Switzerland, where this manuscript, based on the work presented by M. di C. at Epioptics 11 (Erice 2010), could be completed.

128

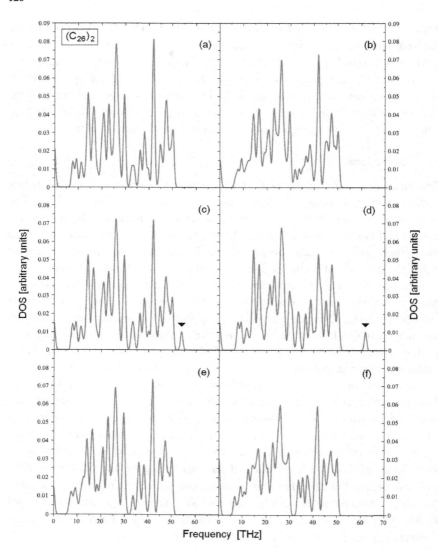

Fig. 8: Same as in Fig. 6 for (C26)2.

## References

[1]    K. S. Novoselov, A. K. Geim, S. V. Morozov, D. Jiang, Y. Zhang, S. V. Dubonos, I. V. Grigorieva, A. A. Firsov, *Science* **30**, 666 (2004).

[2]     K. S. Novoselov, D. Jiang, T. Booth, V. V. Khotkevich, S. M. Morozov, and A. K. Geim, *Proc. Natl. Acad. Sci. U.S.A.* **102**, 10451 (2005).

[3]     K. S. Novoselov, A. K. Geim, S. V. Morozov, D. Jiang1, M. I. Katsnelson, I. V. Grigorieva, S. V. Dubonos, and A. Firsov, *Nature,* **438**, 197 (2005)

[4]     A. K. Geim and K. S. Novoselov, *Nature Mater.* **6**, 183 (2007).

[5]     A. H. Castro Neto, F. Guinea, N. M. R. Peres, K. S. Novoselov, and A. K. Geim, *Rev. Mod. Phys.* **81**,109 (2009).

[6]     H. W. Kroto, J. R. Heath, S. C. O'Brien, R. F. Curl and R. E. Smalley, *Nature* **318**, 162 (1985).

[7]     S. Iijima, *Nature* **324**, 56 (1991).

[8]     D. Donadio, L. Colombo, P. Milani and G. Benedek, *Phys. Rev. Lett.*, **84,** 776 (1999).

[9]     E. Barborini, P. Piseri, P. Milani, G. Benedek, C. Ducati and J. Robertson, *Appl. Phys. Lett.* **81**, 3359 (2002); highlighted by E. Gerstner, *Nature, Materials Update,* 7 Nov 2002 (http://www.nature.com/materials/news/news/021107/portal/ m021 107-1.html)

[10]    G. Benedek, H. Vahedi-Tafreshi, E. Barborini, P. Piseri, P. Milani, C. Ducati and J. Robertson, *Diamond and Rel. Mater.* **12**, 768 (2003).

[11]    A. V. Rode, E. G. Gamaly, A. G. Christy, J. G. Fitz Gerald, S. T. Hyde, R. G.Elliman, B. Luther-Davies, A. I. Veinger, J. Androulakis, and J. Giapintzakis, *Phys. Rev. B* **70**, 054407 (2004); highlighted by R. F. Service, *Science* **304**, 42 (2004).

[12]    D. Arčon, Z. Jagličič, A. Zorko, A. V. Rode, A. G. Christy, N. R. Madsen, E. G. Gamaly, and B. Luther-Davies, *Phys. Rev. B* **74**, 014438 (2006).

[13]    L. Diederich, E. Barborini, P. Piseri, A. Podestà, P. Milani, A. Scheuwli, and R. Gallay, *Appl. Phys. Lett.* **75**, 2662 (1999).

[14]    I. Boscolo, P. Milani, M. Parisotto, G. Benedek, and F. Tazzioli, *J. Appl. Phys.* **87,** 4005 (2000)

[15]    G. Benedek, P. Milani, and V. G. Ralchenko, Eds., *Nanostructured Carbon for Advanced Applications* ((Kluver, Dordrecht 2001) and papers therein.

[16]    A.C. Ferrari, B. S. Satyanarayana, J. Robertson, W. I. Milne, E. Barborini, P. Piseri, and P.Milani, *Europhys. Lett.*, **46**, 245 (1999).

[17]    G. Bongiorno, C. Lenardi, C. Ducati, R.G. Agostino, T. Caruso, M. Amati, M. Blomqvist, E. Barborini, P. Piseri, S. La Rosa, E. Colavita and P. Milani, J. Nanosci. Nanotechnol. **10**, 1 (2005).

130

[18]   S. Agarwal, X. Zhou, F. Ye,  Q. He, G. C. K. Chen, J. Soo, F.    Boey, H. Zhang, and P. Chen, *Langmuir Lett.* **26**, 2244 2010).

[19]   A. L. McKay, *Nature* **314**, 604 (1985).

[20]   A. L. McKay and H. Terrones, *Nature* **352**, 762 (1991).

[21]   H. Terrones and A. L. McKay, in *The Fullerenes*, H. W. Kroto, J. E. Fisher and D. E. Cox, Eds. (Pergamon Press, Oxford 1993) p. 113.

[22]   M. O'Keeffe, G. B. Adams and O. F. Sankey, *Phys. Rev. Lett.* **68**, 2325 (1992).

[23]   T. Lenosky, X. Gonze, M. Teter and V. Elser, *Nature*, **355**, 333 (1992).

[24]   S. J. Townsend, T. Lenosky, D. A. Muller, C.S. Nichols and V. Elser, Phys. Rev. Lett. 69, 921 (1992).

[25]   D. Vanderbilt and J. Tersoff, *Phys. Rev. Lett.* **68**, 511 (1992).

[26]   K. H. A. Schwarz, *Gesammelte Mathematische Abhandlungen* (Springer, Berlin 1890).

[27]   P. Milani e S. Iannotta, *Cluster Beam Synthesis of Nanostructured Materials* (Springer, Berlin 1999).

[28]   M. Bogana, D. Donadio, G. Benedek and L. Colombo, *Europhys. Lett.*, **54**, 72 (2001).

[29]   G. Benedek, H. Vahedi-Tafreshi, P. Milani and A. Podestà, "Fractal Growth of Carbon Schwarzites" in *Complexity, Metastability and Non-Extensivity*, edited by C. Beck et al. (World Scientific, Singapore 2005), p. 146-155.

[30]   G. Benedek, M. Bernasconi, E. Cinquanta, L. D'Alessio, and M. De Corato, in *Mathematics and Topology of Fullerenes*, ed. by F. Cataldo, A. Graovac, and O. Ori, Springer Series on Carbon Materials Chemistry and Physics, Vol. 5 (Springer, Heidelberg Berlin 2011), Chap. 12.

[31]   D. E. Manolopoulos and P. W. Fowler, JCP **96** 7603 (1992).

[32]   I. Làszlò, A. Rassat, P. W. Fowler, and A. Graovac, Chem. Phys. Lett. **342**, 369 (2001).

[33]   L. D'Alessio, Thesis, University of Milano-Bicocca, 2007 (unpublished).

[34]   M. de Corato, M. Bernasconi, L. d'Alessio, O. Ori, and G. Benedek, in *Topological Modeling of Nanostructures and Extended Systems*. ed. by O. Ori et al, Springer Series in        Carbon Materials: Chemistry and Physics, Vol. 5 (Springer,    Heidelberg Berlin 2012), Chap. 8.

[35]   R. L. Cappelletti, J. R. D. Copley, W. A. Kamitakahara, F. Li,    J.S. Lannin, and D. Ramage, Phys. Rev. Lett. **66**, 3261 (1991)

[36]   L. Pintschovius, Rep. Prog. Phys. **59** (1996) 473, and references therein.

[37]   L.A. Chernozatonskii, Physics Letters A 172 (1992) 173.

[38]    S. Spadoni, L. Colombo, P. Milani and G. Benedek, Europhysics Lett. **39** (1997) 269.

[39]    J. M. Romo-Herrera, M. Terrones, H. Terrones, S. Dag, and V. Meunier, Nano Lett. **7** (2007) 570.

[40]    G. Benedek and L. Colombo, in *Cluster Assembled Materials*, ed. by K. Sattler (Trans Tech Publ. Ltd, Zürich 1996) p. 247-274.

[41]    X. Blase, G. Benedek, and M. Bernasconi, in *Computer-Based Modeling of Novel Carbon Systems and their Properties,* ed. by L. Colombo and A. L. Fasolino, Springer Series in Carbon Materials: Chemistry and Physics, Vol 3, (Springer, Berlin Heidelberg 2010), Chap. 6.

[42]    G. Benedek and G.F. Nardelli, J. Chem Phys. 48, 5242 (1968).